This book is to be returned on or before
the last date stamped below.

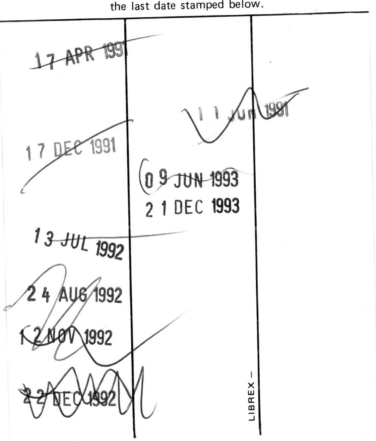

17 APR 1991

17 DEC 1991

13 JUL 1992

24 AUG 1992

12 NOV 1992

22 DEC 1992

11 JUN 1991

09 JUN 1993

21 DEC 1993

LIBREX —

APPLIED VISION
1989 TECHNICAL DIGEST SERIES, VOLUME 16

POSTCONFERENCE EDITION

**Summaries of papers presented at the
Applied Vision Topical Meeting
Topical Meeting**

July 12–14, 1989

San Francisco, California

Cosponsored by the

National Aeronautics and Space Administration
Optical Society of America

Optical Society of America
1816 Jefferson Place, N.W.
Washington, D.C. 20036
(202) 223-8130

ISBN Number
     Conference Edition   1-55752-103-4 (softcover)
     Postconference Edition 1-55752-104-2 (hardcover)
     (Note: Postconference Edition includes postdeadline papers.)
     1989 Technical Digest Series 1-55752-105-0 (hardcover)

Library of Congress Catalog Card Number
     Conference Edition   89-61730
     Postconference Edition 89-61729

TABLE OF CONTENTS

TUESDAY, JULY 11, 1989

MONTEREY ROOM

6:00 PM–9:00 PM   REGISTRATION/RECEPTION

WEDNESDAY, JULY 12, 1989

MONTEREY ROOM

7:30 AM–5:30 PM   REGISTRATION/SPEAKER CHECKIN

EMPIRE ROOM

9:00 AM–10:00 AM
WA   OPENING ADDRESS
James Larimer, *NASA Ames Research Center, Presider*

9:00 AM   (Invited Paper)
WA1   Forty Years of Image Technology and Vision Research: A Personal Memoir, D. H. Kelly, *SRI International.* These memoirs include stories of people involved in aircraft camouflage, motion picture technology, information compression, and visual science, from World War II to the present day. (p. 2)

MONTEREY ROOM

10:00 AM–10:30 AM   COFFEE BREAK

EMPIRE ROOM

10:30 AM–12:00 M
WB   MODELS OF HUMAN VISUAL IMAGE CODING
Walter Makous, *University of Rochester, Presider*

10:30 AM   (Invited Paper)
WB1   Neural Codes, Receptive Fields, and Visual Representations, Andrew B. Watson, *NASA Ames Research Center.* Spatial imagery is represented in the brain in the responses of populations of neurons, or neural codes. The use of digital implementations of these neural codes to evaluate image fidelity is reported. (p. 4)

11:00 AM   (Invited Paper)
WB2   Multiscale Transforms for Image Processing, Edward H. Adelson, Eero P. Simoncelli, *MIT Media Laboratory.* We discuss several multiscale image transforms, including Gabor transforms, Laplacian pyramids, and QMF pyramids, and compare their utility for different tasks in the representation and analysis of image information. (p. 5)

11:30 AM
WB3   Mutlirate Filter Bank Perspective of Retinal Processing, Bennett Levitan, Gershon Buchsbaum, *U. Pennsylvania.* We propose hybrid parallel and hierarchic processing implementations for the retina by considering the computational and connectional complexities of multirate filter banks. (p. 9)

11:45 AM
WB4   Is Fractal Image Compression Related to Cortical Image Compression?, Michael D. McGuire, *Hewlett Packard Laboratories.* Barnsley's fractal image compression technique considered as transformation replacement instead of random iteration algorithm appears remarkably similar to current ideas on cortical image compression. (p. 13)

EMPIRE ROOM

1:30 PM–3:00 PM
WC   SPATIAL AND TEMPORAL IMAGE SAMPLING AND DISPLAY
Albert J. Ahumada, Jr., *NASA Ames Research Center, Presider*

1:30 PM   (Invited Paper)
WC1   Photoreceptor Sampling of Moving Images, David R. Williams, *U. Rochester.* Biological imaging systems can exhibit aliasing artifacts: the perceived direction of motion of interference fringes imaged on the retina is sometimes reversed as a result of aliasing by the human cone mosaic. (p. 18)

2:00 PM   (Invited Paper)
WC2   Vision Research Applied to Computer Synthesized Imagery, Franklin C. Crow, *Xerox PARC.* Vision research has continually influenced computer graphics from early work on the rendering of smooth surfaces to recent work on stochastic sampling techniques. (p. 20)

2:30 PM
WC3   Learning in Interpolation Networks for Irregular Sampling: Some Convergence Properties, Albert J. Ahumada, Jr., Jeffrey B. Mulligan, *NASA Ames Research Center.* Some convergence properties are presented for Ahumada and Yellott's spontaneous-activity-based interpolation learning networks and a version of Maloney's eye-movement-based learning algorithm. (p. 21)

2:45 PM
WC4   Calibrating a Linear Visual System by Comparison of Inputs Across Camera/Eye Movements, Laurence T. Maloney, *New York U.* A method is discussed that geometrically calibrates a vision system, without explicit feedback and without knowing photosensor positions, by learning to compensate for eye movements. (p. 28)

MONTEREY ROOM

3:00 PM–3:30 PM   COFFEE BREAK

EMPIRE ROOM

3:30 PM-5:15 PM
WD   IMAGE COMPRESSION AND COMMUNICATION
William K. Pratt, *Sun Microsystems, Presider*

3:30 PM   (Invited Paper)
WD1   Operators for Facial Feature Extraction, D. E. Pearson, E. Hanna, *U. Essex, U.K.* This study shows that different operators are required for different parts of the face, with a composite operator being best for overall facial feature extraction. (p. 34)

4:00 PM   (Invited Paper)
WD2   Image Processing by Intensity Dependent Spread, Tom N. Cornsweet, John I. Yellott, *UC–Irvine.* A relatively new theory of human visual processing called intensity-dependent spread has some interesting properties as a digital image processing algorithm. (p. 38)

4:30 PM
WD3   Image Characteristics Recovery from Bandpass Filtering, Rachel Alter-Gartenberg, *Old Dominion U.;* Ramkumar Narayanswamy, *Science & Technology Corp.* Images filtered with bandpass filters preserve most of their original characteristics. Quantifying them, we recover the original intensity and reflectance representations from the bandpassed data. (p. 41)

4:45 PM
WD4   Temporal Compression of American Sign Language Using Event Boundaries, David H. Parish, *U. Minnesota;* George Sperling, Michael S. Landy, *New York U.* A temporal compression scheme for low bandwidth transmission of American Sign Language is developed. Compressed signal intelligibility is evaluated and compared with a control scheme. (p. 45)

5:00 PM
WD5   Image Compression in Noise, Scott Daly, *Eastman Kodak Co.* A noise adaptive contrast sensitivity function is used in image data compression. This approach allows the compression algorithm to be optimized for different imaging systems. (p. 49)

MONTEREY ROOM

7:30 AM-5:30 PM   REGISTRATION/SPEAKER CHECKIN

EMPIRE ROOM

8:30 AM-12:00 M
ThA   IMAGE QUALITY METRICS
Carlo Infante, *Digital Equipment Corporation, Presider*

8:30 AM   (Invited Paper)
ThA1   Effective Range of Viewing Instruments, Aart van Meeteren, *TNO Institute for Perception, The Netherlands.* Image quality of viewing instruments can be expressed in an effective retinal pixel element, representing resolution, low contrast, low luminance, and noise effects all together. (p. 54)

9:00 AM   (Invited Paper)
ThA2   Perceptual Image Quality Metrics, Jacques A. J. Roufs, Huib de Ridder, Joyce Westerink, *Instituut v. Perceptle Onderzoek, The Netherlands.* To establish useful relationships between perceptual quality and the physical image parameters, quantitative measures of subjective quality and its underlying dimensions are necessary. (p. 58)

9:30 AM   (Invited Paper)
ThA3   Quality Measures in Digital Halftones, Paul G. Roetling, *Xerox Corp.* Over the last two decades the use of digitally generated halftone images has increased significantly. This paper considers quality measures such as gray levels represented, tone reproduction, and noise. (p. 59)

MONTEREY ROOM

10:00 AM-10:30 AM   COFFEE BREAK

EMPIRE ROOM

10:30 AM
ThA4   Evaluation of Subjective Image Quality with the Square Root Integral Method, Peter G. J. Barten, *Barten Consultancy, The Netherlands.* Experimental data on subjective image quality at varying resolution, addressability, contrast luminance, and display size are compared with predictions by the square root integral. The correlation appears to be good. (p. 63)

10:45 AM
ThA5   Visual Multipoles for Quantifying Raggedness and Sharpness of Images, Stanley Klein, Thom Carney, *UC–Berkeley.* A new formalism using visual multipoles can be used to quantify image quality by specifying edge raggedness and sharpness in terms of multipole moments. (p. 69)

11:00 AM
ThA6 Distortion Metrics for Image Coding Using Monochrome and Color Human Visual Models, Scott E. Budge, *Utah State U.* The use of a human visual model based distortion measure for image coding and evaluation is described. Test results show that if an image is encoded using the visual model, the measured quality correlates better with visual quality when the measurements are made within the model, and better images are produced. (p. 73)

11:15 AM
ThA7 Psychophysical Rating of Image Compression Techniques, Charles S. Stein, Lewis E. Hitchner, *UC–Santa Cruz;* Andrew B. Watson, *NASA Ames Research Center.* Image compression schemes abound with little work which compares their bit-rate performance based on objective perceptual fidelity measures. We used a psychophysical method to estimate, for a number of compression techniques, a threshold bit-rate yielding a criterion level of performance in discriminating original and compressed images. (p. 76)

11:30 AM (Invited Paper)
ThA8 Perceptual Gains for Coding of Moving Images Without Visible Impairments, Bernd Girod, *MIT Media Laboratory.* A new nonlinear spatiotemporal model of human threshold vision is proposed that accurately predicts a variety of perceptual effects. Maximum bit-rate savings by irrelevancy reduction are evaluated for natural test pictures based on rate distortion theory. (p. 81)

11:45 AM
ThA9 Image Gathering and Digital Restoration: End-to-End Optimization for Visual Quality, Friedrich O. Huck, *NASA Langley Research Center;* Sarah John, Judith A. McCormick, Ramkumar Narayanswamy, *Science & Technology Corp.* We demonstrate how the fidelity, resolution, sharpness, and clarity produced by conventional approaches to image gathering and digital restoration can be significantly improved. (p. 85)

EMPIRE ROOM

1:30 PM–3:15 PM
ThB READING AND DISPLAY LEGIBILITY
Curt R. Carlson, *David Sarnoff Research Center, Presider*

1:30 PM (Invited Paper)
ThB1 Reading: Effects of Contrast and Spatial Frequency, Gordon E. Legge, *U. Minnesota.* The effects of contrast reduction and spatial frequency filtering on reading speed can be related to known properties of sensory mechanisms in vision. (p. 90)

2:00 PM (Invited Paper)
ThB2 Why Was Reading Slower from CRT Displays than from Paper?, John D. Gould, *IBM Research Center.* Behavioral experiments demonstrated that, with better resolution displays and with fonts, polarity, and layout resembling those of paper, people read as fast from CRT displays as from paper. (p. 94)

2:30 PM
ThB3 Effects of Character Size and Chromatic Contrast on Reading Performance, Kenneth Knoblauch, Aries Arditi, *The Lighthouse for the Blind.* We studied the influence of chromatic and luminance contrast on reading rate for a range of character sizes. (p. 98)

2:45 PM
ThB4 Reading and Contrast Adaptation, Dennis Pelli, *Syracuse U.* Reading white on black text, but not black on white text, substantially increases the contrast threshold for detecting a sinusoidal grating with the same line frequency as the text. (p. 102)

3:00 PM
ThB5 Image Quality Metric for Digital Letterforms, Joyce E. Farrell, Andrew E. Fitzhugh, *Hewlett Packard Laboratories.* To measure digital letterform quality objectively, font designers need automated tools that predict the visual system's response to the characters displayed on various devices and in various viewing conditions. We describe a metric for predicting the discriminability of digitized representations of high resolution characters presented on low resolution display devices. (p. 104)

MONTEREY ROOM

3:15 PM–3:45 PM   COFFEE BREAK

EMPIRE ROOM

3:45 PM–5:30 PM
ThC HIGH DEFINITION AND EXTENDED DEFINITION TELEVISION
William E. Glenn, *Florida Atlantic University, Presider*

3:45 PM
ThC1 Perception of Jutter in Temporally Sampled Images, William E. Glenn, Karen G. Glenn, *Florida Atlantic U.* Measurements of contrast sensitivity for jutter perception are described as a function of sampling frequency, spatial frequency and velocity for luminance and isolumenant chromaticity gratings. (p. 110)

11:30 AM
FA7   Chromatic Subsampling for Display of Color Images,
Claude Sigel, RuthAnn Abruzzi, James Munson, *Digital Equipment Corp.* We measured the detectability of representing chromatic information at fewer spatial locations than achromatic information in color images. For many pictures, reducing the chromatic information by a factor of 16 or more was not detected. (p. 158)

EMPIRE ROOM

1:00 PM–2:30 PM
FB   COLOR CODING
Gerald Murch, *Tektronix, Presider*

1:00 PM   (Invited Paper)
FB1   Segregation of Basic Colors in an Information Display, Robert M. Boynton, Harvey Smallman, *UC–San Diego.* Will time to locate a small spatial detail hidden among ten each of ten focal basic color targets reveal serial search limited to the attended color? (p. 164)

1:30 PM
FB2   Color and Visual Search in Large and Small Display Fields, Allen L. Nagy, Robert R. Sanchez, Thomas C. Hughes, *Wright State U.* Color differences required for parallel search were measured for small and large display fields and small and large stimuli. Both factors affect results. (p. 168)

1:45 PM
FB3   Theoretical Constraints on the Participation of Rods and Cones in Color Matches, Michael H. Brill, *Science Applications International Corp.* We prove rod–cone spectral independence, identify convergence conditions for iterative tetrachromatic matching, and discuss conditions under which mesopic color matching would require only three primaries. (p. 172)

2:00 PM
FB4   Light Source Size and the Stimulus to Vision, James A. Worthey, *U.S. National Institute of Standards & Technology.* Source size affects the range of color and luminance in a scene. Some new calculations of these phenomena are presented. (p. 176)

2:30 PM
FB5   Detection of Spatial Frequency Selected Color Shifts and the Contrast Sensitivity Functions for CRT Primaries, Hirohisa Yaguchi, Hidemi Takahashi, Yoichi Miyake, *Chiba U., Japan.* The relationship between the contrast sensitivity functions of CRT primary colors and the detectability of spatial frequency selected color shifts is analyzed. (p. 180)

NOTES

NOTES

NOTES

NOTES

NOTES

xiv

NOTES

WEDNESDAY, JULY 12, 1989

9:00 AM–10:00 AM

WA1

OPENING ADDRESS

James Larimer, NASA Ames Research Center, *Presider*

40 Years of Image Technology and Vision Research
A Personal Memoir

D. H. Kelly

SRI International

Menlo Park, CA 94025

From student days to the present, the author reviews his adventures in designing optics, camouflaging airplanes, simulating the atmosphere, bypassing military procedures, going Hollywood, inventing photographic processes, making interference filters, decoding aerial photographs, building visual stimulators, and finally becoming a visual scientist. The narrative includes episodes from the University of Rochester, USC and UCLA, the wartime Navy in Washington, Mitchell Camera and Technicolor Corporations in Hollywood, Itek Corporation in Lexington and Palo Alto, and SRI International. No scientific material will be presented.

WEDNESDAY, JULY 12, 1989

10:30 AM–12:00 M

WB1–WB4

MODELS OF HUMAN VISUAL IMAGE CODING

Walter Makous, University of Rochester, *Presider*

Neural Codes, Receptive Fields, and Visual Represenations

Andrew B. Watson
NASA Ames Research Center
Moffett Field, CA 94035

Spatial imagery is represented in the brain in the responses of populations of neurons, or neural codes. The use of digital implementations of these neural codes to evaluate image fidelity will be discussed.

Hexagonal QMF pyramids.

Edward H. Adelson[†] and Eero P. Simoncelli[‡]

Media Laboratory
[†]Department of Brain and Cognitive Science
[‡]Department of Electrical Engineering and Computer Science
Massachusetts Institute of Technology
Cambridge, Massachusetts 02139

It is widely recognized that effective image processing and machine vision must involve the use of information at multiple scales, and that models of human vision must be multi-scale as well. The most commonly used image representations are linear transforms, in which an image is decomposed into a sum of elementary basis functions. Besides being well understood, linear transformations in the form of convolutions provide a useful model of some of the early processing in the human visual system.

The following properties are valuable for linear transforms that are to be used in image processing and vision modeling:

Completeness (Invertibility): Invertibility guarantees that a transformation does not discard any information about the image. In the standard linear algebra terminology, the basis set of an invertible transformation is said to be *complete*. **Localization in Scale (Radial SF):** As noted above, another property that is considered to be important in image processing is an explicit representation of *scale*. As an equivalent description of the notion of scale, we may refer to sub-bands in the frequency domain representation of the image. **Orientation tuning (Angular SF):** Orientation tuning is an important property of most cortical cells and psychophysically inferred channels. In machine vision it is important in the detection of lines and edges. **Spatial localization:** It is often important for a representation to contain locational information; thus the basis functions should have compact support. This may also improve computational efficiency. **Overlap:** The above requirements can only be satisfied with basis functions that overlap smoothly in space and spatial frequency. When a transform uses sharp-edged blocks, the result is poor frequency localization and block artifacts. Similarly, it is desirable for the frequency responses of the different bands to blend smoothly into each other. **Orthogonality (Self-inversion):** In many cases it is advantageous to work with representations that are orthogonal, or more generally, self-inverting. A self-inverting transform is one whose basis functions and inverse or *sampling* functions are identical. In the case of a linearly independent basis set, this condition is equivalent to orthogonality. **Computational Efficiency:** A final property worth considering is that of computational efficiency. There are two ways to achieve efficiency in convolution-based linear transforms. We can use basis functions which have a small region of support, or we can use basis functions which are computable as cascades of simple operations. Both are typically used in pyramids.

We have developed a transform that captures all of the above properties; it is based on quadrature mirror filters (QMF's) with hexagonal symmetry. QMF's are by definition orthogonal, and we have described methods for designing QMF's that are well localized in space and spatial frequency [1]. Most research with two dimensional QMF's has involved separable filters, in which case at least one of the subbands contains mixed orientations [2, 3, 4], or non-separable filters which are not orientation-specific [2]. We have previously described a hexagonal QMF pyramid in which all bands are oriented [5]. In the present work we apply such a pyramid to several tasks. We also note that the same concept can be extended to three dimensions; if the third dimension is taken to be time, the resulting filters will be tuned for motion. A more detailed discussion may be found in [1].

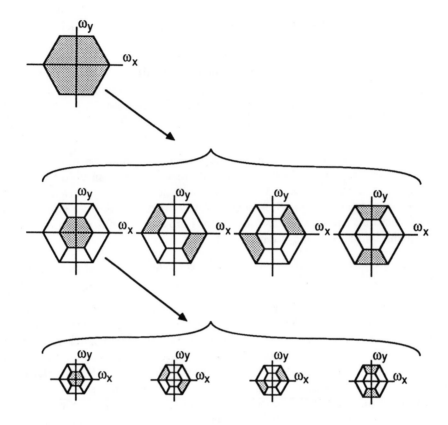

Figure 1: Idealized diagram of the partition of the frequency plane resulting from a four-level pyramid cascade of hexagonal filters. The top plot represents the frequency spectrum of the original image. This is divided into four sub-bands at the next level. On each subsequent level, the lowpass sub-band (outlined in bold) is sub-divided further.

Figure 1 illustrates the way the hex QMF pyramid works in the frequency domain. The image is sampled on a hexagonal grid, and its spectrum is assumed to be limited to a hexagonal region of the Fourier plane. At each stage the pyramid subdivides the spectrum into four subbands; one is low-pass and the other three are oriented and band-pass. This process is repeated with subsampling by a factor of 4 at each level. The resulting pyramid constitutes a self-similar orthogonal image transform. The application of the hex QMF pyramid to a disc image is shown in figure 2 as described in the figure caption.

We applied the pyramid to the task of image data compression, as shown in figure 3. The pyramid coefficients were quantized and entropy coded, to reduce the data rate from 8 bit/pixel to 1 bit/pixel.

Because all of the hex filters are oriented, they offer a useful set of measurements of local image properties. In figure 4 we illustrate their use in analyzing local orientation of an image. Oriented energy measures were taken over the Einstein image, and these were used to derive the both the strength and angle of the local orientation.

The concepts used in the design of the hexagonal QMF's may be extended to three di-

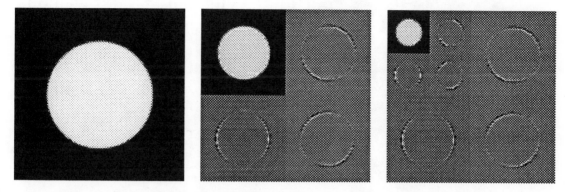

Figure 2: Results of applying a hexagonal QMF bank to an image of a disk. On the left is the original image. In the center is the result after one application of the analysis section of the filter bank. The image has been decomposed into a low-pass and three oriented high-pass images at 1/4 density. On the right, we have applied the filter bank recursively to the low-pass image to produce a two-level pyramid decomposition.

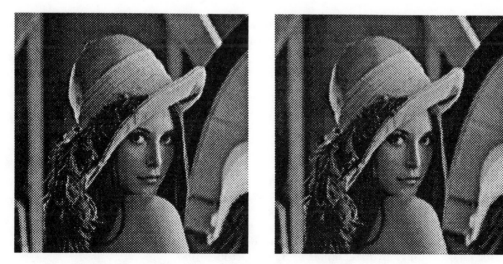

Figure 3: Data compression using the 4-ring hexagonal filter bank. On the left is the original "Lena" image at 256×256 pixels. On the right is a the compressed image. The entropy of the quantized pyramid was 1.0 bits per pixel for a total of 65536 bits.

Figure 4: On the left is the original image of Albert Einstein. On the right is an oriented line drawing produced by using linear combinations of the oriented hexagonal filter outputs to measure the strength and orientation of the local image anisotropy.

mensions. Just as a hexagonal lattice is tightest packing of disks in two dimensions, a rhombic dodecaheral lattice is the tightest packing of spheres in three dimensions. It is possible to design three dimensional QMF pyramids, in which the basis functions are tuned for various three dimensional orientations. If the third dimension is time, the filters can be used for measuring local motion energy [1].

References

[1] Eero P. Simoncelli and Edward H. Adelson. *Non-separable Extensions of Quadrature Mirror Filters to Multiple Dimensions.* Vision Science Technical Report 119, Vision Science Group, Media Laboratory, Massachusetts Institute of Technology, 1989. In press, Proceedings of the IEEE (1990).

[2] Martin Vetterli. Multi-dimensional sub-band coding: some theory and algorithms. *Signal Processing*, 6(2):97–112, February 1984.

[3] John W. Woods and Sean D. O'Neil. Subband coding of images. *IEEE Trans. ASSP*, ASSP-34(5):1278–1288, October 1986.

[4] S. G. Mallat. *A theory for multiresolution signal decomposition: the wavelet representation.* GRASP Lab Technical Memo MS-CIS-87-22, University of Pennsylvania, Department of Computer and Information Science, 1987.

[5] Edward H. Adelson, Eero Simoncelli, and Rajesh Hingorani. Orthogonal pyramid transforms for image coding. In *Proceedings of SPIE*, October 1987.

A Multirate Filter Bank Perspective of Retinal Processing

Bennett Levitan and Gershon Buchsbaum
Department of Bioengineering
School of Engineering and Applied Science
University of Pennsylvania
220 South 33rd St., Towne Building
Philadelphia, PA 19104-6392

I. ABSTRACT

A multirate filter bank (MRB) consists of a set of filters that produce several reduced sampling-rate versions of the same input signal (fig. 1). They have been found useful for subband coding [1, 2], image compression [3] and other areas [3, 4]. The possible implementations for MRBs tradeoff computational with connectional complexities and handle processor noise differently. We explore the consequences of the various MRB implementations for retinal processing and show that fully parallel processing or hybrid parallel/hierarchic processing are candidates for the retina.

II A. Multirate filter bank (MRB) view of the retina.

Psychophysical evidence indicates that the retina can be viewed as an MRB. Wilson and Bergen [5] and Sekuler, et. al., [6], show that a four mechanism model is required to account for human threshold responses to spatially-localized, aperiodic patterns and to sinusoidal gratings. Each mechanism has a center-surround spatial weighting function described as a difference of gaussians (DOG) (fig. 2, from Wilson and Bergen [5]). All mechanisms operate in parallel at all points of the image.

The temporal classification and the distribution of retinal receptive field sizes provide physiologic evidence supporting a retinal MRB implementation like that of Wilson and Bergen. For example, the smaller-sized tuned sustained mechanisms, N and S in fig. 2, may correspond to X-cells in the cat and to color-opponent cells in the primate [7]. The larger size-tuned, transient mechanisms, T and S in fig. 2, may correspond to Y-cells in the cat and to spectrally-broadband cells in the primate.

II B. MRB implementations tradeoffs

The number of possible implementations of MRBs depends on the number of outputs, N, and the frequency-domain properties of the filters [8]. Implementation are defined by three parameters: the hierarchic index, H_I; the parallel index, P_I; and the serial index, S_I.

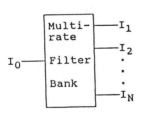

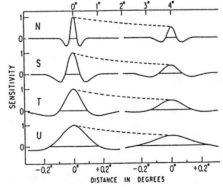

Fig. 1. Generic multirate filter bank. The bank produces N output images by filtering and reducing the sampling rate of image I_0.

Fig. 2. Weighting functions of the four mechanisms from Wilson and Bergen's model. The diagram shows their normalized spatial weighting functions at 0° and their variation as a function of eccentricity to 4°. The upper scale indicates eccentricity; the lower scale shows distance relative to the center of each mechanism. (from Wilson and Bergen [5])

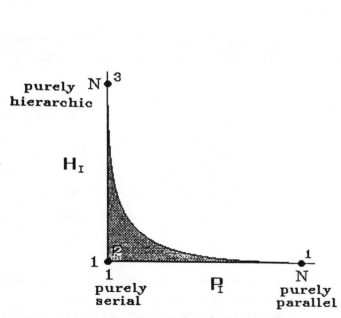

$H_I = 2$ (a)
$P_I = 1$
$S_I = 1$

$H_I = 3$ (b)
$P_I = 2$
$S_I = 1$

$H_I = 1$ (c)
$P_I = 2$
$S_I = 2$

complexity	MRB Implementation		
	fully hierarchic	fully parallel	fully serial
computational connective	highest lowest	lowest highest	highest highest

Fig. 3. Sample multirate filter bank implementions. The symbols are explained in the text.

Table 1. Relative computational and connective complexities in the extreme MRB implementations.

The hierarchic index is the maximum number of filters in one cascade in the MRB. The parallel index is the maximum number of cascades that operate simultaneously. Fig. 3 shows examples of MRBs in which the filters are represented by vertical arrows. Each filter operates either on the original image, indicated by a large dot, or on the output of another filter, indicated by an indexed letter "O", O#. For example, fig. 3a displays an MRB in which the input image is filtered to produce image O1 which in turn is filtered to produce image O2. Since the cascade consists of two filters, $H_I = 2$. There is only one cascade, so $P_I = 1$. Fig. 3b shows an MRB with $P_I = 2$ and $H_I = 3$. Finally, MRBs may have several sets of cascades that operate in succession. The maximum number of consecutive cascades is S_I. Fig. 3c shows an MRB with $S_I = 2$.

From the examples in fig. 3, it follows that the number of outputs, N, equals $H_I \cdot P_I \cdot S_I$ where $1 \leq H_I$, P_I, $S_I \leq N$. This equation defines a 3-dimensional surface on which all implementations of an MRB may be found. In fig. 4, we show the "implementation diagram," a projection of this surface onto the H_I-P_I plane. MRBs are found only in the shaded region and on its boundary. The edge points labeled 1, 2 and 3 designate MRB implementations that are purely parallel, purely serial and purely hierarchic respectively. All other points correspond to implementations that are mixtures of these three properties.

Filters are often rated according to how rapidly they operate and the hardware resources they require. For each MRB implementation, we measure a computational and a connectional complexity. We define computational complexity as $S_I \cdot H_I$, the maximum total number of outputs produced by a succession of cascades. This

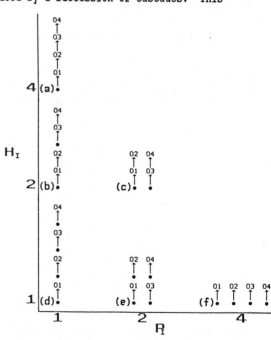

Fig. 4. Implementation diagram for multirate filter bank with arbitrary number of outputs, N.

Fig. 5. Implementation diagram for multirate filter bank with 4 outputs. Particular implementations (a) - (f) are placed approximately where they lay in the space.

accurately reflect the time for computation; since, in an MRB, all the multiplications and additions in a filter occur simultaneously. For example, this measure is 2 for fig.s 3a and 3c and is 3 for fig. 3b. We define connectional complexity as the total number of coefficients used in the filters in an MRB. This number provides a measure of the hardware's complexity. Table 1 shows the relative complexities for the three extreme implementations. Hierarchic processors, because they partially compute their upper level outputs by computing their lower level outputs, have smaller connectional complexity than the other implementations. The tradeoff in the complexities of intermediate MRB implementations is not simple and depends greatly on the extent of the MRB filters. The general nature of the tradeoff is as shown in fig. 6a [9]. Reducing H_I in the vicinity of N substantially lowers the computational complexity but only slightly increases the connectional complexity. Only when H_I approaches 1 does the connectional complexity rises rapidly. Depending on how one weighs the two complexity measures, an optimal MRB implementation can be found on the curve. Another important implementation-dependent filter property is resistance to the influence of processor noise. In a fully parallel implementation, an error in one output has no effect on other outputs. In an hierarchic processor, the error in lower level outputs may propagate into higher levels and accumulate.

II C. MRB implementations in the retina

We consider the retina as an MRB with four outputs. The implementation diagram for N = 4 shows all possible MRB implementations (fig. 5). MRB implementation 5d is purely serial ($S_I = 4$). Implementations 5b and 5e are partially serial.($S_I = 2$) and partially hierarchic and parallel. Implementations 5a, 5c and 5f are as minimally serial as possible ($S_I = 1$) with different combinations of parallel and hierarchic processing. In principle, all six implementations are mathematically viable candidates for an MRB [8]. Which of these six implementations are feasible for the retina?

Unquestionably, all cells in the retina operate continuously and simultaneously. This requires that $S_I = 1$ and eliminates implementations 5b, 5d and 5e. The remaining three implementations are members of a continuum between fully hierarchic and fully parallel implementations. We use two approaches to explore which may be used by the retina.

Insufficient anatomical information exists for generating the entire complexity tradeoff curve for the retina. However, insight can be gained from an example curve for the N = 4 case. Fig. 6b is typical of complexity tradeoff curves for the $S_I = 1$, N = 4 case. Connective complexity corresponds to the number of anatomical pathways between cells in the retina. Computational complexity corresponds to the time lag between light hitting the photoreceptors and a signal appearing on the optic nerve. The two extreme implementations on the curve in fig. 6b have relatively inferior computational or connective complexity. The middle implementation would appear a good compromise. However, since we do not know how these complexities are weighted in the retina nor the exact form of the complexity tradeoff curve, this inference must be taken cautiously.

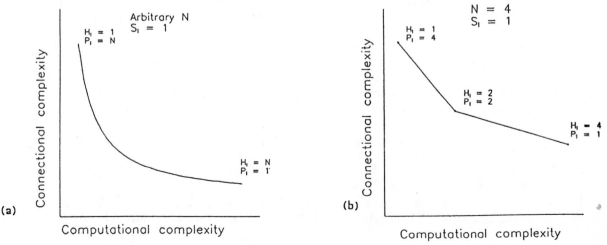

Fig. 6. Typical computational and connectional complexity tradeoff curves for multirate filter banks with (a) arbitrary number of outputs, N; (b) Four outputs.

11

The implementations also differ greatly in their handling of retinal noise. Because of the propagation and accumulation of errors in an hierarchy, it is likely that the final two outputs of implementation 5a will be corrupt with noise. This is a strong argument against implementation 5a. While a similar problem could occur with implementation 5c, we feel that one hierarchic stage is unlikely to corrupt the signal severely.

III. DISCUSSION

Parallel processing concepts are often attributed to the retina. The notion of purely hierarchic implementations of retinal receptive fields has been discussed as well [10, 11]. In this paper, using the formalization of MRBs, we present retinal implementation possibilities which use hierarchic and hybrid parallel/hierarchic architectures. These implementations vary in their computational and connectional complexity depending on the degree of hierarchic and parallel processing.

An outstanding issue is the exact physiological correlate of the MRB implementation and the its location in retina or cortex. Wilson and Bergen indicate that their model may reflect cortical as well as retinal processing [5]. While it is difficult to infer the exact sites of processing from psychophysical studies and currently available anatomical data, the following points favor retinal processing. 1. The temporal classification of retinal receptive fields into sustained and transient classes agrees well with Wilson and Bergen's four mechanisms. 2. The profile of retinal center-surround receptive fields, which is well-fit by DOGs, indicates that the necessary hardware to implement the mechanisms exists in the retina. 3. We believe the distribution of receptive field radii [7] is broad enough to allow for the anatomic variation of several size-tuned mechanisms.

IV. REFERENCES

1. J. W. Woods and S. D. O'Neil, "Subband coding of images," *IEEE Trans. on ASSP*, Vol. ASSP-34, No. 5, 1278-1288, 1986
2. R. E. Crochiere, S. A. Webber and J. L. Flanagan, "Digital Coding of Speech in sub-bands," *Bell Systems Tech. Journal*, Vol. 55, No. 8, 1976
3. A. Rosenfeld, (Ed),<u>Multiresolution Image Processing and Analysis</u>, Springer-Verlag, Berlin, 1984
4. D. Marr, <u>Vision: A computational investigation into the human representation and processing of visual information</u>, Freeman, San Francisco (1982)
5. H. R. Wilson and J. R. Bergen, "A four mechanism model for threshold spatial vision," *Vision Research*, Vol. 19, 19-32, 1979
6. R. Sekuler, H. R. Wilson and C. Owsley, "Structural modeling of spatial vision," *Vision Research*, Vol. 24, 713-719, 1984
7. M. De Monasterio and P. Gouras, "Functional properties of ganglion cells of the rhesus monkey retina," *Journal of Physiology*, Vol. 251, 167-195, 1975
8. B. Levitan and G. Buchsbaum, "An Hierarchic Processing Scheme for Arbitrary Multirate Filter Banks," SPIE 1989 Visual Communications and Image Processing Conference
9. B. Levitan and G. Buchsbaum, "Properties of multirate filter bank implementations," in preparation
10. J. Burt and E. Adelson, "The laplacian pyramid as a compact image code," *IEEE Trans. on Communications*, Vol. COM-31, No. 4, 532-540, 1983
11. G. Hartmann, "Recursive features of circular receptive fields," *Biological Cybernetics*, Vol. 43, 199-208, 1982
12. E. Zrenner, <u>Neurophysiological Aspects of Color Vision in Primates (Studies of Brain Function</u>), Springer-Verlag, Berlin, Heidelberg, 1983

Is Fractal Image Compression Related to Cortical Image Compression?

Michael D. McGuire
Hewlett Packard Laboratories
1501 Page Mill Rd.
Palo Alto, CA 94304

Fractals are customarily introduced with simple line replacement rules, for example the well known Koch curve. At each state of iteration single lines are replaced by combinations of lines according to a rule. Remarkable complexity can be built up this way as shown by five stages of this line replacement bush fractal[1].

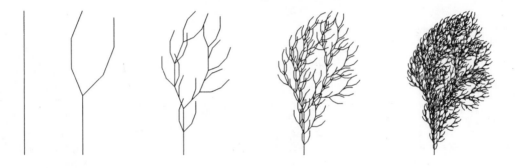

But a more general approach to generating deterministic fractals is that of Barnsley et al. [2,3,4] which involves <u>transformation</u> replacement. The transformations are linear affine transformations. Applied to a figure they can move, rotate, scale, shear, or invert it. If the figure is a square, most generally an affine transformation makes it a parallelogram moved and rotated from its original position. The "R-square" below demonstrates this.

The mathematical form of an affine transformation is

$$X_{new} = X_{old}\ r \cos(x, x') - Y_{old}\ s \sin(y, y') + X_{offset}$$

$$Y_{new} = X_{old}\ r\ \sin(x, x') + Y_{old}\ s \cos(y, y') + Y_{offset}.$$

Where r is the scaling of the x to x' axes, s is the scaling of the y to y' axes, (x, x') is the angle between the old and transformed x axes, and (y, y') is the angle between the old and transformed y axes, and x_{offset} and y_{offset} are as shown. If r = s = 1 , xoffset = yoffset = 0, and (x, x') = (y, y'), the transformation is a simple rotation.

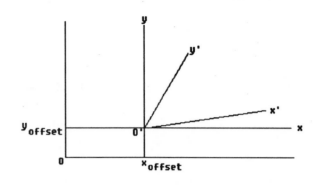

Fractals can be generated by iterating the application of a set of several downsizing or contracting transformations to an input image--a simple example: three transformations which halve the size of the input image and position themselves with respect to the original as shown.

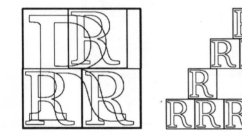

Each iteration is to replace the previous image with all the transformations of it. Here a Sierpinski triangle is emerging. Of course any input shape would have worked including a simple point. The "R" and the square show the effect of the transformations. Consider the following set of four transformations, demonstated with the square--one of them is the inobvious short vertical line at the upper left.

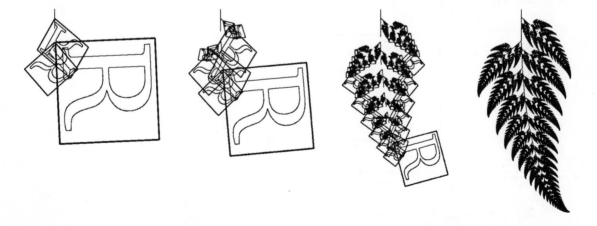

Iteration shrinks the squares onto this fern shape. It is the <u>attractor</u> of iteration of this set of transformations. Barnsley refers to this as an <u>iterated function system</u>.

14

This is in the context of a fast algorithm he discovered for obtaining the attractor from the transformations. The method is iterative and selects randomly from the set of transformations. It is described in detail elsewhere[3]. However the transformation replacement approach is the heuristically more useful in this discussion.

Transformations are deduced and images compressed using Barnsley's collage theorem. An image is segmented in some convenient way--e.g. gray scale or color. For each segment an outline is found. The outline is more or less covered or tiled with transformed copies of itself--a mouse or other pointer controlled computer graphics environment is a nice way to do it. There is a certain amount of art to it. The transformations that produce the copies in this way are what is wanted.

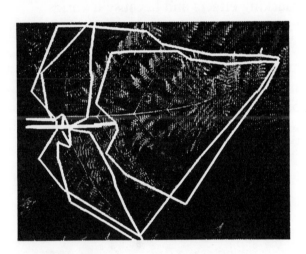

There are ideas about how images are represented and decorrelated in the visual cortex which involve self-similarity and sensitivities to angle, position and spatial frequency exemplified in work by Watson and Ahumada[5], Field[6], and Daugman[7]. There is some intersection with the transformation replacement approach to fractals to explore. Watson is the most specific algorithmically and thus the most convenient to discuss. They proposed a hexagonal orthogonal-oriented pyramid which decorrelates and compresses information from the approximately hexagonally close-packed cells of the retina. Their pyramid can be naturally generated using transformation replacement starting with the highest level hexagon, and then superposing the iterations.

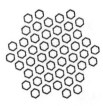

The smallest hexes represent the retinal input. Each of the second level hexes organizes the seven smallest ones at its center and corners in seven different ways as three edge detectors, three grating detectors and a blob detetector. Each of the edge and grating detectors are angle sensitive to one of three different orientations to cover all angles and are band pass in spatial frequency. The blob detector is low pass and forms the contribution to the third level hex which organizes it and the contributions from other six nearest second level hexes as the second level organized the first and so on up the pyramid. The responses of the detectors form the coding of an image which comes to them. The number of responses is of the same magnitude as the number of input pixels. But the amount of information per response that would have to be kept for perceptually lossless reconstruction of the image can be reduced because of contrast masking effects and because it varies with spatial frequency which varies with level in the pyramid.

This coding process then appears to be doing the inverse of transformation replacement but with the bandpass edge and grating information being preserved in each level. Consider the reconstruction of an image that is the pyramid itself. There would be a single point at the peak and the transformation replacement series would be as below.

It lacks the lines and edges because the information was intentionally eliminated. The difference then is that the fractal approach directly encodes and passes all information to the "top of the pyramid" leaving nothing at lower levels. Whether it is in practice a more efficient coding will depend on how much segmentation is needed to "fractalize" an image. The proliferation of sets of transformations needed to represent an image may defeat its efficiency.

1. D. Saupe, (1988) A Unified Approach to Fractal Curves and Plants, in The Science of Fractal Images, H. O. Peitgen and D. Saupe (eds.) Springer Verlag, New York, pp 274--286.
2. M.F. Barnsley, V. Ervin, D. Hardin, and J Lancaster, (1976) Proc. Nat. Acad. Sci. U.S.A. 83, 1975-1977.
3. M. F. Barnsley and A. D. Sloan (1988), Byte Magazine , January, 215--223.
4. M. F. Barnsley, (1988) Fractals Everywhere, Academic Press, San Diego, Ca.
5. A. B. Watson and A. J. Ahumada, (1989) IEEE Trans. Biomed. Eng. 36, 97--105
6. D. J. Field, (1987) J. Opt. Soc. A. A4, 2379--2394
7. J. G. Daugman, (1989) IEEE Trans. Biomed. Eng. 36, 107--114

WEDNESDAY, JULY 12, 1989

1:30 PM–3:00 PM

WC1–WC4

SPATIAL AND TEMPORAL IMAGE SAMPLING AND DISPLAY

Albert J. Ahumada, NASA Ames Research Center, *Presider*

PHOTORECEPTOR SAMPLING OF MOVING IMAGES

David R. Williams
Center for Visual Science
University of Rochester, Rochester NY 14627

SUMMARY

Biological imaging systems can exhibit aliasing artifacts: the perceived direction of motion of interference fringes imaged on the retina is sometimes reversed as a result of aliasing by the human cone mosaic.

ABSTRACT

A periodic moving stimulus can appear to move in the reverse direction if it is undersampled in time, as in the case of the "wagon wheel" effect caused by an inadequate frame rate in motion pictures. Sampling by a spatial array of sensors or pixels can produce a similar motion reversal for periodic patterns moving at any velocity, if the spatial sampling frequency is too low. These artifacts are well-known to engineers who design discrete imaging systems. The artifact resulting from spatial undersampling has been demonstrated in biological imaging systems (Goetz, 1965, Coletta and Williams, 1987). For example, insects tethered at the center of a rotating drum containing low spatial frequency vertical stripes exhibit an optomotor response: they rotate in the same direction as the stripes. However, these insects reverse their direction of motion when confronted with spatial frequencies that exceed the Nyquist frequency of their ommatidial array. This is just what one would expect from spatial aliasing by the regular array of insect ommatidia. Nancy Coletta and I have demonstrated a similar effect in the human with drifting interference fringes whose contrast is immune to optical degradation. In the parafoveal retina, high spatial frequency (but not low) gratings look like two-dimensional spatial noise and can appear to move in the *opposite* direction from their true direction of motion. This motion reversal can be demonstrated with a forced-choice technique. Subjects guessed the direction of motion of vertical, unity contrast fringes whose direction was randomly determined on each trial. No feedback was provided. Percent correct falls significantly below chance performance at high spatial frequencies, indicating a reversal in the perceived direction of motion. At higher frequencies, the perceived direction of motion reverses a second time, and at even higher frequencies performance settles to chance.

It is of some interest to know how the frequencies at which these motion reversal occur relate to the spatial grain of the visual system at each eccentricity. A simplified model that invokes spatial sampling to explain the motion reversals might include a one-dimensional regular sampling array followed by a spatial filter that passes spatial frequencies up to the Nyquist frequency and no higher (sinc interpolation). In such a model, the perceived direction of motion would reverse at integer multiples of the Nyquist frequency. However, the visual system samples the retinal image with a disordered, two-dimensional photoreceptor mosaic. Disorder in the sampling array causes aliasing energy to be dispersed into a broad range of spatial frequencies, orientations, and velocities. Carlo Tiana and I have constructed a more realistic model in which gratings are sampled by cone mosaics taken from the monkey retina. The model incorporates blurring by the cone aperture, which ultimately limits the highest frequencies at which motion in any direction can be observed. Following sampling by the mosaic, a spatial filter is applied that weights the spatial frequency components that ultimately feed the final stage of the model in which a decision about direction of motion is made. The model shows that disorder actually found in primate mosaics has only a modest effect on the spatial frequencies producing the motion reversals. The second motion reversal, in particular, is stable with large changes in the model parameters, indicating that the motion reversal phenomenon can estimate photoreceptor spacing in the living eye.

Goetz K.G. (1965) Behavioral analysis of the visual system of the fruitfly Drosophila. In *Proceedings of the Symposium on Information Processing in Sight Sensory Systems,* pp. 85-100. California Insitute of Technology, Pasadena, CA.

Coletta N.J. and Williams D.R. (1987b) Undersampling by cones reverses perceived direction of motion. *Invest. Ophthal. Vis. Sci. 28, 232.*

Supported by AFOSR-85-0019, EY04367, EY01319.

Vision Research Applied to Computer-Synthesized Imagery

Franklin C. Crow
Xerox PARC
Palo Alto, California

Summary

Vision Research and Computer Graphics

Computer Graphics has always drawn on vision research although often rather indirectly. From the beginning, issues such as flicker fusion and resolution had to be considered in designing systems for computer graphics. However, there was a large body of experience and well-established standards from the cinema and television industries to provide guidelines. Later, when research focused more on the attainment of realism in images computed from numeric models, issues were raised which required a more basic understanding of vision. Currently, while the community, as a whole, is aware of the importance of underlying principles of vision, there is little reference to vision research in journals and conferences on computer graphics. On the other hand, many advances in computer graphics were based on results found in the vision literature. A few examples follow.

Smooth Shading

Around 1970 researchers at the University of Utah were concerned with achieving a smooth appearance to the polygonal surfaces they were depicting. Even when smooth surfaces were approximated by large number of polygons, the surfaces still looked polygonal (Figure 1a). With some direction from Tom Stockham, Henri Gouraud learned that the apparent frequency response of the retina works to enhance small discontinuities in intensity, the Mach band effect. Gouraud referenced Ratliff [Ratliff65] as his source.

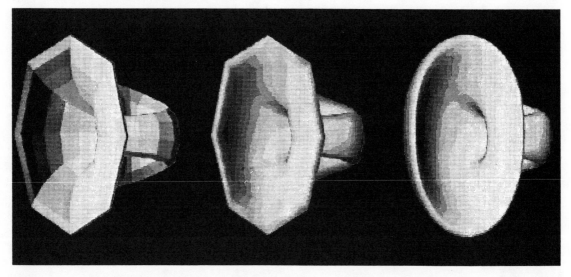

Figure 1: (a) A Klein bottle approximated by polygons, (b) with smooth shading using Gouraud's method, (c) using bicubic patches instead of polygons.

This led Gouraud to suggest an interpolation scheme which guaranteed continuity across polygon boundaries on a surface [Gouraud71]. Instead of computing a color for each polygon, compute a color at each vertex of the polygonal mesh. Colors at interior points can then be found by bilinear interpolation. This worked quite well (Figure 1b) and the method is still in wide use today.

Antialiasing

The author [Crow78] and others [Naiman87] have found support in the vision literature (they reference Cornsweet [Cornsweet70]) for the notion that size can be traded for intensity in very small objects (one to two minutes of arc and less) and, similarly, that shapes are indistinguishable at such sizes. Broadcast standards for television are predicated on similar measurements . The quantizations forced by the regular array of pixels on standard digital displays cause jagged edges and other artifacts if it is assumed that details can be no smaller than the size of a pixel. By variations in intensity the position and apparent size of small details can be varied and the artifacts can be diminished.

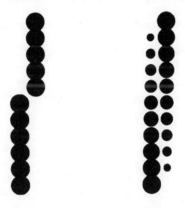

Figure 2: (a) A near vertical line, approximated by spots at pixel locations. (b) The same line represented more accurately by varying spot sizes.

As an example consider a nearly vertical line drawn on a raster (Figure 2a). The rectangular grid imposed by a conventional display forces a jump from one column of pixels to the next every now and then if the positions followed by the line are restricted to pixel locations. By using variations in intensity the apparent position of the line can be placed more precisely (Figure 2b), thereby eliminating the sudden jumps that are otherwise so distracting. To ensure the success of the technique, the display must be calibrated so that two half brightened pixels emit the same summed energy that a single fully brightened pixel does. In other words the output of the display must be linear.

How many Grey Levels?

There has been a very practical concern all along in the computer graphics community with the number of grey levels required on a display. Additional grey levels require more complex equipment to display and more space to store. Naturally, there is always pressure to minimize anything of that nature. When making images of smoothly shaded shapes using Gouraud's method (described above) or other techniques, the number of grey levels becomes quite important. An insufficient number of grey levels causes unsightly bands to appear over the surface (Figure 3). Similar problems occur in natural images including large areas of sky or other smoothly varying phenomena.

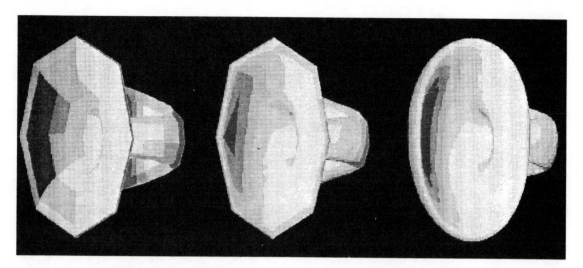

Figure 3: The image from figure 1 restricted to 8 grey levels (3 bits per pixel).

Drawing on early contrast discrimination data from Blackwell [Blackwell46], the number of grey levels necessary can be computed from the available contrast ratio. Blackwell showed that inability to discriminate brightness levels appears at something around a 2% variation in intensity. Therefore the number of discriminable grey levels can be related to a contrast ratio as follows:

$$C = 1.02^n \qquad (1),$$

where C is the contrast ratio and n is the number of grey levels.

Factors enforced by the nature of computer hardware make either 16 or 256 grey levels (corresponding to 4 or 8 bits per pixel) very convenient. Given 16 grey levels, eq. 1 says that a contrast ratio of only 1.37 : 1 can be expected, not very interesting. On the other hand 256 levels allows a contrast ratio as high as 159 : 1, sufficient for all but the most extreme situations. Those with demanding applications sometimes use 4096 grey levels (12 bits per pixel), allowing contrast ratios exceeding 10^{35}, a bit beyond what is necessary.

Interestingly, the above assumes an exponential distribution of intensities where the needs of anti-aliasing require a linear distribution. This causes a problem in displays with only 256 grey levels as adjacent darker shades can often be discriminated on a linear display. Although this does not appear to be difficult to overcome, I have not seen a published implementation of a solution to this problem.

Sampling Distributions

Some more recent work in computer graphics has been concerned with avoiding aliasing artifacts due not only to spatial quantization but also due to temporal quantization. This led to the use of stochastic sampling in spacetime to determine the color of a pixel [Cook86]. Given the screen area represented by a pixel and the time interval represented by a frame, a set of rays can be fired from the synthetic eyepoint through the pixel area at the numerically described scene as defined during a given time interval. The color returned by the surface hit by each ray can be averaged with those of other rays to get an approximation of the color that would be seen by an imaging device aimed at some physical realization of the scene.

This is a very powerful and increasingly popular technique. However, the question of what sampling pattern to use is immediately important. Cook looked into the vision literature and noted Yellott's evidence for a Poisson disc distribution of photoreceptors in the less densely packed areas of the retina [Yellott83]. A similar sampling distribution has proven to work well and some subsequent work has focused on finding more efficient ways to implement such a distribution in practical image generation systems.

Future Possibilities?

Thus far, perhaps the largest effort in computer graphics research has gone into producing the most realistic images possible. With the help of vision research it may be possible to go beyond that goal into a much more useful realm. Computer graphics has been best at representing what can't otherwise be seen: imaginary environments, the visual impact of proposed designs, physical phenomena too small or too vast to photograph, abstractions which help us to understand complex concepts, etc. Furthermore, as computer graphics becomes less expensive, it is increasingly becoming a medium for illustration, replacing more limited tools such as pen and ink and the airbrush. The goal of computer graphics should really be to produce the most *understandable* images possible. To approach such a goal requires a better grasp of what makes an image understandable.

References

[Blackwell46] H. R. Blackwell, "Contrast Thresholds of the Human Eye", *JOSA*, v. 36, pp 642-643, (1946)

[Cook86] Robert L. Cook, "Stochastic Sampling in Computer Graphics", *ACM Transactions on Graphics*, **5**, 1, pp. 51-72, (1986).

[Cornsweet70] T. N. Cornsweet, *Visual Perception*, Academic Press, New York, 1970

[Crow78] Franklin C. Crow, "The Use of Grayscale for Improved Raster Display of Vectors and Characters", Proc. Siggraph '78, *Computer Graphics*, **12**, 2, pp. 1-5, (1978).

[Gouraud71] H. Gouraud, "Computer display of curved surfaces", *IEEE Trans. Comput.*, **C-20**, (1971), 623-629.

[Naiman87] Avi Naiman and Alain Fournier, "Rectangular Convolution for Fast Filtering of Characters", Proc. Siggraph '87, *Computer Graphics*, **21**, 4, pp. 233-242 (1987).

[Ratliff65] F. Ratliff, *Mach Bands: Quantitative Studies on Neural Networks in the Retina*, (Holden-Day, San Francisco, 1965).

[Yellott83] J. I. Yellott, Jr., "Spectral Consequences of Photoreceptor Sampling in the Rhesus Retina", *Science*, 221, July 22, 1983, pp. 382-385.

Learning in Interpolation Networks for Irregular Sampling: Some Convergence Properties

Albert J. Ahumada, Jr. and Jeffrey B. Mulligan
NASA Ames Research Center
Mail Stop 239-3
Moffett Field, California 94035

Introduction

Recently, Ahumada and Yellott (1) and Maloney (5,6) have presented schemes for training networks designed to reconstruct irregularly sampled retinal images. In these schemes adjustable weighting networks provide compensation for the irregularities in the retinal array and the geometrical distortions in intermediate pathways. This paper presents some ideas relating to the convergence of the training algorithms.

Position Learning

The Ahumada and Yellott method for generating an interpolation network is composed of two separate procedures. First the arrangement of the receptors is learned by the higher level; then interpolation weights are learned. The receptor positions are transferred to the higher level because the interpolation learning has a feedback loop which includes the receptor positions, and there is presumably no appropriate path from higher centers back to the retina. In the position learning algorithm the jth receptor has a position vector r_j and an associated position at time n, x_{jn}, at the higher level. Let i index the receptor activated at time n. The active receptor causes the others' associated positions to move so that the distances from it are the same at both levels. That is,

$$x_{jn+1} = x_{in} + (x_{jn}-x_{in})|r_j-r_i|/|x_{jn}-x_{in}|.$$

This rule is equivalent to the learning rule described by Ahumada and Yellott if the learning is carried out to asymptote for each activation. Figure 1 illustrates the motion of one point under the influence of two others.

The resulting learning process can be thought of as having two phases. First the correct neighbor relationships are established and then the distances are refined. Once the errors in the distances are much smaller than the distances, it can be shown that all adjustments reduce the total squared error. Figure 1 also shows what can happen for three points in the small error case. Each adjustment zeroes the error component in the direction of the active point. If adjusting points for two consecutive times are in orthogonal directions from a point, the error for that point is essentially eliminated. If the angle is small, the change in the error will be small.

Unfortunately, the correct configuration may not be obtained in the first place. Figure 2 shows a configuration which does not converge if the activation proceeds cyclically. If activations are random, eventually enough activations would all be at one end and the system would configure correctly. Fortunately, this type of problem does not seem possible with the compact arrangements of receptors with which we are concerned.

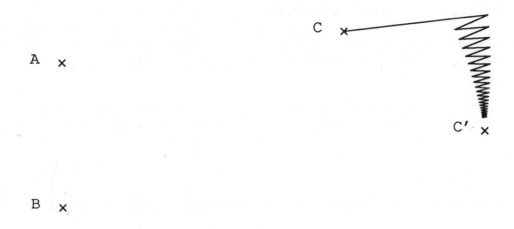

Figure 1. Path of a point C under alternate activation of points A and B. Points A and B are correctly positioned according to their retinal positions, but C begins away from C', which corresponds to its retinal position. As A and B are alternately activated, the position of C moves so that the distance from C to the active point is correct. Asymptotically, each activation reduces the error by the multiplicative factor of the cosine of the angle AC'B.

Figure 2. A non-converging configuration is obtained when the retinal points corresponding to A, B, C, and D are in the configuration A, B, D, and C, respectively. The lines indicate the point motions as the activation proceeds in the order A, B, C, D. The path of D begins towards C as in Figure 1, but the activation of C sets it back. The aspect ratio of the rectangle is 3:1.

Interpolation weight learning

After the receptor positions are available at the higher level, Ahumada and Yellott propose that a network W is trained to transform the input image so that a following lowpass filter stage results in an interpolation of the image. The lowpass filters define a spread matrix S whose rows s_i are the impulse responses at each of the receptor positions. The interpolation will be correct at the sample positions if W is the inverse of S, that is,

$$WS = I.$$

The weight learning mechanism activates the spread matrix rows one at a time, feeds them back into the compensation network and adjusts the network weights by the delta rule so that the output of the compensation network is the corresponding row of the identity matrix. That is,

$$W_{n+1} = W_n - k s_i^T e_{in},$$

where W_n is the weight matrix for time n, k is a scalar learning rate, and e_{in} is the row vector of errors,

$$e_{in} = s_i W_n - u_i,$$

where u_i is the ith row of the identity matrix.

The delta (LMS or Widrow-Hoff) rule convergence properties have been extensively studied (2,4,7,8). The convergence behavior of the delta rule is easy to understand in the case that k is set equal to $1/s_i s_i^T$. Then each iteration zeroes the component of every e_{jn} in the direction of s_i. In the case of two receptors, the behavior of the error vectors is exactly like the errors plotted in Figure 1. The angle between the points A and B is now the angle between the two smoothing function vectors. Again, if the spread functions are orthogonal, the errors will be eliminated after each receptor unit has been activated once, but if the receptor units are highly correlated, the convergence can be very slow. There are, of course, well known schemes for increasing the convergence rate of the delta rule (3). Ahumada and Yellott describe two methods for this special case which are convenient for sequential processing. Only in the case that the correlations are relatively small does the method make sense biologically, because otherwise the precision of the calculations becomes critical.

Learning interpolations from motion

Maloney has developed an elegant way of learning interpolation functions from image translations. In the version we consider here, the input image is represented by a row vector s_i and the effect of translating the image is assumed to be representable as a matrix transformation T. For regular arrays, translations by integral numbers of samples are represented by permutation matrices. Internally, the image is transformed by the compensation matrix W_n to $s_i W_n$. It is further assumed that the internal representation of the image translation can be computed by another transformation t. If W_n does not correctly represent the structure of the original array, there will be an error in the difference between the internally translated image $s_i W_n t$ and the internal representation of the externally translated image, $s_i T W_n$,

$$e_{in} = s_i T W_n - s_i W_n t.$$

We train the weights using an equation which looks like the delta rule except that the quantity playing the role of the desired response, $s_i W_n t$, is now a variable:

$$W_{n+1} = W_n - k (s_i T)^T e_{in}.$$

The behavior of this rule is easiest to understand in the simple case where the receptor array and the internal array are both regular and the motion is an integer number of sample steps, so that both $\mathbf{T}$ and $\mathbf{t}$ can be represented by the same permutation matrix. Also we will assume that the input images are impulses (rows of the identity matrix). For this case, if $k=1$, the adjustment rule above copies the translated row of the weight matrix corresponding to the impulse over the row corresponding to the tranlated impulse. For appropriate sequences of impulses and motions, the adjustment rule can generate a translation invariant filter whose impulse response is one lucky row of $\mathbf{W}_0$. The process can converge in as few iterations as the number of receptors, but if the positions of the impulses and the translations are chosen at random, the time to converge can be quite long. For $k=1/2$, the new position gets an impulse which is the average of its current value and the one from the original position, so that the final impulse response tends to an average of all the impulse responses when positions are chosen at random. Also, if one impulse response is fixed, the rest will eventually converge to it.

Investigations of the adjustment procedure along the lines described above for the delta rule show that convergence is speeded up by having the input images be orthogonal to each other and to their translations, as they are in the simple impulsive case above. Also, the analysis for regular sampling can be extended to irregular sampling by arguments similar to those of Stone (7) which show that training an irregular input is related to training the regularly sampled input by a correlation matrix computed from the transformation relating the two samplings.

Conclusion

Although both of the above methods compute compensation matrices, their goals differ. The interpolation network attempts to make the interpolated image correct at the receptor positions, while the motion invariance network tries to make the impulse responses position-independent. For both methods, orthogonality of training stimuli and structure in the presentation schedule are important determinants of the rate of convergence.

References

1. Ahumada, A. J., Jr. & Yellott, J. I., Jr., Reconstructing Irregularly sampled images by neural networks. *Proc. SPIE Conf.#1077*, Paper#27 (1989).
2. Bitmead, R. R., Persistence of exitation conditions and the convergence of adaptive schemes. *IEEE Trans. Info. Th.*, **30**, 183-191 (1984).
3. Jacobs, Robert A., Increased rates of convergence through learning rate adaptation. *Neural Networks*, **1**, 295-307 (1988).
4. Kohonen, T., *Self-organization and associative memory*. Berlin: Springer-Verlag (1984).
5. Maloney, L. T., Spatially irregular sampling in combination with rigid movements of the sampling array. *Inv. Ophth. & Vis. Sci.*, **29**, ARVO Suppl., 58 (1988).
6. Maloney, L. T., Learning algorithm that calibrates a simple visual system. *Opt. Soc. Amer. Tech. Dig. Series*, **11**, 133 (1988).
7. Stone, G. O., An analysis of the delta rule and the learning of statistical associations. in Rumelhart, D. E. & McClelland, J. L. eds., *Parallel Distributed Processing*, vol. I, Cambridge, MA: MIT Press, Chapt. 11, 444-459 (1986).
8. Widrow, B. and Stearns, S. D. *Adaptive signal processing*. Englewood Cliffs, NJ: Prentice-Hall (1985).

CALIBRATING A LINEAR VISUAL SYSTEM BY COMPARISION OF INPUTS ACROSS CAMERA/EYE MOVEMENTS

Laurence T. Maloney

Department of Psychology and Center for Neural Science
New York University

A visual system is calibrated geometrically if its estimates of the spatial properties of a scene are accurate: straight lines are judged straight, angles are correctly estimated, collinear line segments are perceived to fall on a common line. A visual system can fail to be calibrated because of a mismatch between its optics and later visual processing: calibration of computer vision systems typically requires remapping the sensor inputs to compensate for spherical aberration in the camera lens[1].

It's likely that no biological visual system is ever perfectly calibrated, but considerable evidence exists that biological visual systems do compensate for optical distortions and initial neural disorder[2]. Previous work in visual neural development suggests a variety of sources of information that drive calibration[3], and there are computational models of visual neural development based on these cues[4]. Yet, although biological visual systems are known to require patterned visual stimulation to achieve normal organization[5], none of these models requires such stimulation to function, nor do these models address the problem of calibrating to the optics of a visual system.

This paper describes a new calibration method for a model visual system whose photosensors are initially at unknown locations. The method can compensate for optical distortions equivalent to remapping of sensor locations; it depends critically on patterned visual input. The key idea of the method is that 'prewired' transformations in a visual system that correspond to or cancel the effects of camera/eye movements can be used to organize a visual system.

A MODEL LINEAR VISUAL SYSTEM

The model visual system has N photosensors arranged at locations x_i, y_i, $i = 1, N$ in the unit square. The coordinates x_i, y_i are not required to fall on any regular lattice. A *sensor array* is a set of such (x_i, y_i) pairs. The *light image* is a function from the unit square to the non-negative reals, regarded as the mean intensity of light at each location. Light images can be treated as functions in a linear function space[6]. The output of a sensor is the value of the light image at the location

assigned to the sensor. The mapping from light image to sensor array is assumed to be a linear. The output (measured intensity) from the i'th sensor is denoted ρ_i. In vector notation, the instantaneous input from the sensor array is $\rho = [\rho_1,....\rho_N]^T$ called a *sampling code*. If L is a light image, then $\rho(L)$ will denote the sampling code corresponding to L.

Suppose we have a visual system calibrated to take input from an *ideal sensor array* of N sensors at specified, known locations x'_i, y'_i. This sensor array may be a square or hexagonal grid of sensors, but it need not be. The input from the i'th sensor in this ideal array is denoted μ_i. In vector notation, the input from the ideal array is denoted $\mu = [\mu_1,....\mu_N]^T$. $\mu(L)$ will denote the sampling code derived by the ideal array for the light image L, and $\mu(L) : L \rightarrow R^N$ is a linear transformation.

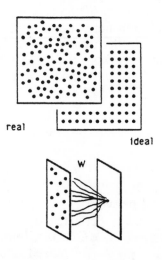

Figure 1: An irregular sampling array can be translated to an ideal regular array.

The visual system receives inputs ρ from its sensors in the real sensor array. It is calibrated for inputs from the ideal array whose input is μ. The calibration problem can be reduced to the following problem for this visual system: learn how to translate from any input sampling code ρ to the corresponding code μ on the ideal array, and perform this translation on its inputs. Figure 1 contrasts the real and ideal sampling arrays. The two-layer visual system shown in Figure 1 is the model visual system.

Without some restriction on the light images sampled, there need be, of course, no connection between samples taken by the real array and those taken by the ideal array unless the positions of some or all of the sensors in the two arrays coincide. For the remainder of this paper, the set of light images **L** is assumed to span a particular lowpass linear function subspace of finite dimension N denoted L^N where N is the number of sensors in the ideal and in the real arrays. This *subspace assumption* is commonly made in modeling spatial vision.

With this assumption, we can show that, if there is a solution to the calibration problem for a particular real array, ideal array, and linear subspace of lights, then it must be a linear transformation. For the remainder of the paper such a (linear) transformation W is assumed to exist; otherwise, calibration is not possible. The problem of calibration is then reduced to learning the linear transformation W, given only input from the real sampling array.

THE CALIBRATION METHOD

Figure 1 shows a receptive field connecting the sensors of the real array to one of the ideal sensors; this receptive field is linear and corresponds to a row of W in matrix form. Figure 2 outlines a procedure for learning W. ρ and ρ' indicate the sampling codes of the real array before and after a transformation T induced by camera movements that effectively translate, 'zoom,' or rotate the light image on the sensor array. T, like W, is a linear transformation[7].

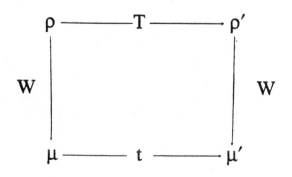

Figure 2: Schematic diagram of the calibration method.

μ and μ' represent the corresponding sampling codes of the ideal array. As noted above, $\mu = W\rho$ and $\mu' = W\rho' = WT\rho$.

It is assumed that there are prespecified internal transformations t which, in a properly calibrated visual system, are equivalent to T in the sense that $tW = WT$. The class of transformations t are precisely the transformations needed to compensate for the consequences of movements. Another way to think of t is that, if the ideal array were used to sample the image before and after an camera movement, then t would be the linear transformation induced by the camera movement on the sampling code μ. The set of transformations t are easily computed and are assumed to be built into the visual system. If the visual system is perfectly calibrated, then the transformations t can mimic the effect of camera/eye movements on the ideal array (translation, 'zoom,' rotation), or, equivalently, can compensate for the effect of eye movements on the ideal array. If the visual system is miscalibrated (W is incorrect), then the class of transformations t will not be able, in general, to mimic or compensate for camera/eye movements.

To simplify presentation of the procedure, the discussion is broken into two cases. *Case 1: perfect knowledge of camera movements.* Suppose, first, that we have perfect control over the direction the camera points, i.e. camera movement is known. Then, in terms of Figure 2, for any input ρ, we can compute two quantities, $tW(\rho)$ and $WT(\rho)$, which are equal when the visual system is calibrated. In terms of the example we began with, we look at the internal image $W(\rho)$, predict what it will look like after a specific camera motion, $tW(\rho)$, perform that camera motion and look at the consequences, $WT(\rho)$. If $tW = WT$ for every pattern ρ and every camera motion T, then the visual system is calibrated.

Otherwise, we form the penalty term $||WT(\rho) - tW(\rho)||^2$ by squaring the discrepancy between the observed and predicted values at each location of the ideal sampling array. Note that this requires that we briefly save the previous visual input for comparision with the current visual input. Recalling that $WT(\rho)$ and $tW(\rho)$ are vectors in $\mathbf{R}^N$, the expression above is the Euclidean distance between the observed and expected vectors. Given a training set of M images with sampling codes ρ^1, ..., ρ^M viewed under camera movements of known extent, we can form an overall penalty term, summed across camera movements and images:

$$\text{Pen}(W) = \sum_{\rho} ||WT(\rho) - tW(\rho)||^2. \quad (1)$$

The goal of the method is to minimize this penalty term **Pen**(W) by choice of W. Under many choices of training set and camera movements, W is uniquely determined by minimization of the quadratic penalty in Eq. (1). The assertion of consistency between the prespecified transformation t and the physical transformations T permits calibration.

Case 2: camera movements approximately known. In any realistic visual system, information about camera direction is imperfectly known, possibly computed in part from the (not yet calibrated) visual input, or itself in need of calibration. The transform t can be chosen, for example, not from prior knowledge of camera motion, but so as to minimize the term $||WT(\rho) - tW(\rho)||^2$: in effect, we assume the transformation that 'best explains' what happened to the visual input.

IMPLEMENTATION

Eq. (1) above can be minimized by any of a number of standard numerical methods including adaptive or 'neural' network algorithms.

Fig. 3, for example shows an irregular sampling array (solid squares) and one sensor of the ideal array (open square). a conjugate gradients method[8] was used to minimize the quadratic penalty term in Eq. (1) and compute the receptive field for the one ideal sensor. The light images used in training were finite Fourier series. The resulting receptive field is correct to machine precision. Repeating the computation for the other ideal sensors in Fig. 1 would produce a calibrated visual system.

Implementations of the method using conjugate gradients and the LMS rule[9] (an adaptive network algorithm) correctly computed receptive fields for regular and irregular real sampling arrays. The LMS rule is tolerant of small errors in camera/eye movement information (See Ref. [7] for details).

EXPERIMENTAL PREDICTIONS

The method described here uses a novel cue, derived from comparision of visual input across successive fixations, to calibrate a simple linear visual system. Previous models of visual neural development, reviewed above, use different cues and methods to calibrate the visual system. The method outlined here differs from these methods in that (a) it directly optimizes a visual capability (stability and rigidity under change of direction of gaze), (b) it can compensate for small optical distortions and remappings, (c) it requires structured visual input (actual scenes), and (d) it requires successive fixations on a single unchanging scene.

Taken as a claim about visual development, the procedures developed here are readily testable empirically. (1) Animals reared in environments lacking structured visual input, or in environments where visual input is rendered perpetually non-rigid, or where it is never possible to fixate the same scene twice, should be perfectly calibrated according to previous theories, but not according to the work developed here. (2) Animals with small optical distortions induced early in development should not be perfectly calibrated according to previous theories, but may be so according to the work described here. (3) The visual system will compensate for retinally-stabilized optical distortions in adults; these distortions may include small induced scotomas. Prediction (2) may hold while (3) fails because recalibration in the adult is likely limited by connectivity restrictions on receptive fields[4].

The last class of experiments can be performed with human observers. The kinds of tasks that would indicate miscalibration and recalibration under any of the above experimental manipulations include: (a) alignment of two line segments by method of adjustment, (b) judging whether the extension of a line segment passes to one or the other side of a target, and (c) judgments of acute/obtuse angles. Forced-choice measures of discrimination are not necessarily affected by miscalibration. In task (a), for example, the observer may discriminate the two stimuli but label the physically misaligned stimulus as aligned.

ACKNOWLEDGEMENTS

Part of this work was presented previously at the 1988 meeting of the Association for Research in Vision and Ophthalmology, and the 1988 meeting of the Optical Society of America. I thank Al Ahumada, Josef Miller, Tony Movshon, Misha Pavel, and Brian Wandell for useful comments on earlier presentations of this work. I especially thank Marty Banks and Jack Yellott for directing my attention toward issues of calibration and performance in vision.

FOOTNOTES

[1] Rosenfeld, A., & Kak, A. C., *Digital Picture Processing*. New York: Academic Press, 1976.

[2] Banks, M.S., Visual recalibration and the development of contrast and optical flow perception. In A. Yonas (Ed.), *Perceptual Development in Infancy; The Minnesota Symposia on Child Psychology, Volume 20*. Hillsdale, New Jersey: Erlbaum, 1988. Wallach, H., *On Perception*. Quadrangle, New York, 1976.

[3] Meyer, R. L., Activity, chemoaffinity, and competition: factors in the formation of the retinotectal map. In Hilfer, S.R., & Sheffield, J. B. (Eds.), *Cell Interactions in Visual Development*. New York: Springer-Verlag, 1988. Purves, D., & Lichtman, J. W., *Principles of Neural Development*. Sunderland, Massachusetts: Sinauer, 1985, pp. 251-270. Shatz, C. J., The role of function in the prenatal development of retinogeniculate connections. In Bentivoglio, M., and Spreafico, R. (Eds.), *Cellular Thalamic Mechanisms*. Elsevier Science, 1988.

[4] von der Malsburg, C., and Willshaw, D., How to label nerve cells so that they can interconnect in an ordered fashion. *Proceedings of the National Academy of Science, USA* **74**, 5176-5178, 1977. Fraser, S. E., A differential adhesion approach to the patterning of nerve connections *Developmental Biology* **79**, 453-464, 1980. Whitelaw, V. A., and Cowan, J. D., Specificity and plasticity of retinotectal connections: a computational model. *Journal of Neuroscience*, **1**, 1369-1387, 1981. Ahumada, A. H. & Yellott, Jr, J. I., A connectionist model for learning receptor positions *Investigative Ophthalmology and Visual Science, Supplement*, **29**, 58, 1988; Ahumada, A. H. & Yellott, Jr, J. I., Training networks to compensate for irregular sampling *Optics News*, **14**, 201, 1988.

[5] Movshon, J. A., & Van Sluyters, R. C., Visual neural development. *Annual Review of Neuroscience*, **7**, Palo Alto, California: Annual Reviews, Inc., 1981.

[6] A linear function space is a vector space whose vectors are themselves functions, here functions from $\mathbf{R}^2$ to $\mathbf{R}$. For details and examples, see Apostol, T. M., *Calculus, Volume 2, 2nd Ed.* Waltham, Massachusetts: Xerox, 1969, Chap 1.

[7] Maloney, L. T., The consequences of discrete retinal sampling for vision. in preparation, 1989.

[8] Press, W. H., Flannery, B. P., Teulosky, S. A., & Vetterling, W. T., *Numerical Recipes; The Art of Scientific Computing*. Cambridge, England: Cambridge University Press, 1986.

[9] Widrow, G., & Hoff, M.E., Adaptive switching circuits. *Institute of Radio Engineers, Western Electronic Show and Convention, Convention Record, Part 4, 96-104, 1960*.

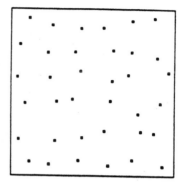

```
 0.0121 -0.0028  0.0435  0.0105 -0.0146  0.0069
-0.0330  0.0659 -0.1819 -0.0930  0.1253 -0.0054
 0.0393 -0.1044  0.3443  0.4181 -0.2920  0.0024
 0.0270 -0.3004  0.6580  0.4219 -0.3062  0.1096
-0.0002  0.0196 -0.0608 -0.0278  0.1890 -0.0517
 0.0021 -0.0161  0.0168  0.0202 -0.0375 -0.0044
```

Figure 3: An irregular array (solid squares) and a single ideal sensor (open square).

WEDNESDAY, JULY 12, 1989

3:30 PM–5:15 PM

WD1–WD5

IMAGE COMPRESSION AND COMMUNICATION

William K. Pratt, Sun Microsystems, *Presider*

OPERATORS FOR FACIAL FEATURE EXTRACTION

D.E. Pearson and E. Hanna

Department of Electronic Systems Engineering
University of Essex
Colchester C04 3SQ
England

1. Introduction

The extraction of human facial features from images in binary form has been shown to be useful in at least two areas of low data-rate image coding. One is in the transmission of moving cartoons over the public switched telephone network [1]-[3]. Another is in model-based coding, where analysis of the camera signal is needed to line up the software model with the exterior world [4]. Binarization of the input image is also used as a first stage in machine recognition of faces [5].

In our research laboratory we are interested in futuristic visual communication systems of the type illustrated in Fig. 1, which bring together various strands of thinking. In such systems, differences between the "motion" of the interior software model and that of the real world occur at a highly variable bit rate, but there is evidence that such data can be transmitted extremely efficiently over packet-switched networks [6].

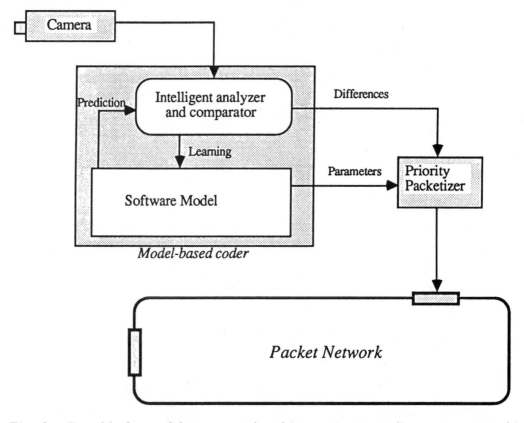

Fig. 1   Possible form of future very low bit-rate image coding system, in which differences between a 3D model and the real world are transmitted at a variable bit rate over a packet-switched network.

2. Object-related feature extraction

Complete systems of the type illustrated in Fig. 1 have not yet been implemented, but several of the component parts have been studied. In particular, we concern ourselves in this paper with the contents of the analyzer box, as it might function when the camera is looking at a human face. In previous work we have found that excellent binary extraction of facial features is accomplished by a composite operator with three component parts: valley detection, edge detection and thresholding [1]. Fig. 2 shows an example, together with a cartoon drawn by a human artist from the same input, namely a grey-level photograph.

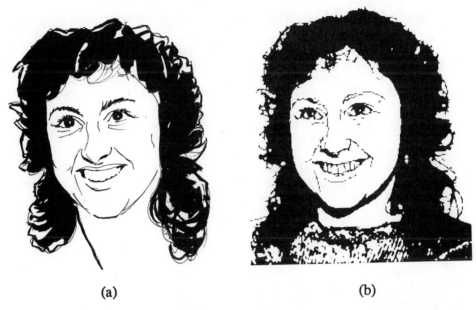

(a)   (b)

Fig. 2. (a) Artist-drawn and (b) operator-extracted cartoon representations of a female subject (from reference [7]).

There is, in general, a complex relationship between 3D object space and the 2D image space in which operators function. We have studied this relationship experimentally by taking simple objects such as spheres, cylinders and cubes and illuminating them at various angles against both dark and light backgrounds. With a camera trained on these objects, we subjected the camera output to valley detection, edge detection and thresholding. The results are represented pictorially in Fig 3.

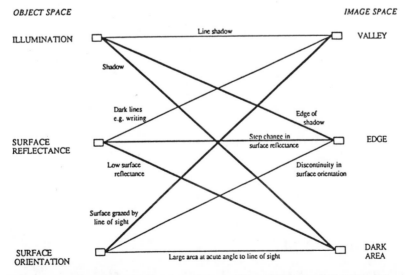

Fig. 3. Relationship between illumination-surface characteristics in object space and the three features in image space which were the subject of the reported investigation. The thicker lines show the restricted set for human faces under common forms of illumination.

Valleys, edges and dark areas (which are detected by thresholding) can each be caused by illumination, surface reflectance or surface orientation. However, in human faces under common forms of illumination, the possibilities are typically reduced to five, as shown by the thicker lines in Fig. 3. One of these, for example, represents the fact that valleys are commonly due to the orientation of the skin surface being at a small angle to the line of sight of the camera, as at the bounding contour of the nose or cheek, or at wrinkles. Our interest in valley detection stemmed originally from a postulate that perceptually important human features occurred at surfaces which were grazed by straight lines drawn from the camera [1].

3. Experimental identification of facial features by human subjects

The idea that different operators are required for different parts of the human face was further explored in experiments involving 26 subjects. In these experiments Harmon's 21-feature facial classification system [8] was used. This system uses 4 descriptors for the hair, 1 for the forehead, and so on. Each descriptor is rated using a category scale.

In our experiments we used 5 different representations of two different faces, seen only in frontal view. One was the face in Fig. 2 and the other was that of a male subject. The representations were:
(i)   an original grey-level photograph,
(ii)   a computer-generated cartoon, as in Fig 2(b), using a composite operator which incorporated valley detection, edge detection and thresholding,
(iii) a "valledge" detector comprising valley and edge detectors only,
(iv)   a threshold detector alone,
(v)   an artist-drawn representation, as in Fig 2(a).

We added a category of "don't know" to each of the category scales, to allow for the limited information presented to our subjects as compared with those of Harmon. The difference between the incidence of "don't knows" for the binary and the grey-level images is indicated in Table 1 below, where negative entries indicate an inferior (more don't knows) and positive entries a superior (fewer don't knows) performance by subjects, as compared with the original grey-level photographs. The table entries are given as a percentage of the total number of responses (shown in column 3), which was the same for each of the last four columns.

Feature	Descriptors	Responses	Full cartoon	Valledges only	Threshold only	Artist-drawn
Hair	4	104	-2	-25	-4	-5
Forehead	1	26	+15	0	+19	+4
Eyebrows	2	52	0	-2	-6	+4
Eyes	3	78	-4	-15	-4	0
Ears	2	52	+6	-17	-2	+6
Cheeks	1	26	-8	-12	-27	0
Nose	3	78	+1	-1	-21	+5
Mouth	4	104	+1	-6	-39	+3
Chin	1	26	+27	-12	+15	+8
TOTAL	21	546	+2	-11	-12	+2

Table 1. Percentage differences (shown in the last four columns) between the number of "don't know" responses for feature-extracted binary versions and the ordinary grey-level version of the same face, when subjects were asked to identify facial features on Harmon's scales.

4. Discussion and Conclusions

Table 1 is a partial presentation of the results of the experiment. The last row of the table indicates that both the full computer-generated cartoon and the artist-drawn cartoon provide as much facial-recognition information as the original grey-level photograph (though the portrayal is not necessarily accurate in the cartoons). By comparison, valledge detection and thresholding by themselves give impoverished representations for recognition purposes. Valledge detection is poor in portraying the hair texture and shade, while thresholding tends to fail in the region of the nose, cheeks and mouth, particularly (in terms of the detailed results), for estimates of the nose length and profile, the 3D shape of the cheeks and the lip thickness and overlap. It has to be noted, however, that both the male and female photographic subjects were smiling, in which action they were showing their teeth but not the shadowy interior cavity of the mouth; other experiments [9],[10] have shown that thresholding is very effective in portraying open mouth shape for speech-reading purposes.

In general, the results confirm that the interplay of light with the complex three-dimensional form and surface-reflectance changes of the human face requires different operators for different parts of the face, with a composite operator being best for overall facial-feature extraction.

5. References

[1]   Pearson D.E. and Robinson J.A., "Visual Communication at Very Low Data Rates", *Proc. IEEE*, vol. 73, no. 4, pp. 795-811, 1985.

[2]   Sperling G., Landy, M., Cohen Y and Pavel M., "Intelligible Encoding of ASL Image Sequences at Extremely Low Information Rates", *Computer Vision, Graphics and Image Processing*, vol. 31, pp. 335-391, 1985.

[3]   Whybray M.W. and Hanna E., "A DSP based videophone for the hearing-impaired using valledge processed pictures", *Proc. IEEE International Conference of Acoustics, Speech and Signal Processing (ICASSP.89)*, paper 41.M9.16, Glasgow, Scotland, 23-26 May, 1989.

[4] Welsh W.J., "Model-based coding of videotelephone images using an analysis/synthesis method", Paper 4.5, *Abstracts of the 1988 Picture Coding Symposium*, Turin, Italy, September 1988.

[5]   Bruce V., *Recognising Faces*, Lawrence Erlbaum Associates, London, 1988.

[6]   Ghanbari M.,   "Two-layer coding of video signals for VBR networks", *IEEE J. on Selected Areas in Communications*, Issue on Packet Speech and Video, to be published, June 1989.

[7] Pearson D.E., Hanna E. and Martinez K., "Computer-generated Cartoons", in *Images and Understanding*, ed. by Horace Barlow and Colin Blakemore, Cambridge University Press, to be published, 1989.

[8]   Harmon L., "The Recognition of Faces", in *Image, Object and Illusion*, Readings from Scientific American, W.H. Freeman & Co., San Francisco, pp. 101-112, 1974.

[9]   Levitt, H., City University of New York, private communication, 1988.

[10]   Brooke N.M., "Looking at speech: studying the analysis, synthesis and perception of visible speech gestures", in Salenieks, (ed.) *Computing: the Next Generation*, Ellis Horwood, Chichester, England, 1988.

Image Processing by Intensity-Dependent Spread

Tom N. Cornsweet and John I. Yellott

Abstract: A relatively new theory of human visual processing, called Intensity-Dependent Spread (IDS), has some interesting properties as a digital image processing algorithm.

Summary

Introduction As the ambient light level changes, many of the properties of human vision also change. For example, if a test flash is smaller than some critical size, its detectability depends only on the total amount of light it delivers, not on the spatial distribution of that light, but when its size exceeds the critical size, its detectability depends only on its flux per unit area. (That is called Ricco's Law.) This critical size is an inverse function of the ambient light level, so that in dim light, the visual system exhibits poorer spatial resolution.

A few years ago, we published a theory that was initially intended to explain the changes in Ricco's Law with ambient light level, but which turned out to manifest a broad range of other phenomena that have been observed in human brightness vision (1,2). Many of these phenomena seem also to be of value in machine vision applications. In this talk, we will first describe the theory and then demonstrate the results of its application in digital image processing.

The Theory

Assume a two-dimensional array of photodetectors, or input pixels, and a parallel two-dimensional array of output pixels, separated by a layer containing a network through which signals from the detectors can spread laterally and reach the output array. When an input pixel is illuminated, a pattern of signals is generated that is centered on the pixel and spreads over a region with some particular profile, eg. Gaussian. We will call this the Signal Spread Distribution (SSD). The amplitude of the SSD is some power function of the input irradiance, and for these purposes we will take it to be a power of one, that is, the peak height of the SSD is linear with irradiance. Further, assume that the signal leaving each output pixel is simply the sum of all the excitation reaching it.

So far, the theory is exactly the standard way of expressing image formation, the SSD corresponding to the point spread function and the output being the convolution of the geometrical image and the point spread function. However, in that standard, linear, model, the total volume of the SSD is linear with irradiance. (The point spread function is defined as the response to an input of <u>unit</u> irradiance; we use the term "Signal Spread Distribution (SSD)" to refer to the pattern resulting from the actual input irradiance.) In the model we are presenting, although the peak amplitude of the SSD is, say, linear with irradiance, its width varies inversely with irradiance in such a way that its volume is constant. That is, we assume that the total amount of signal output by each detector is not changed by changes in irradiance. Instead, the effect of a change in irradiance is to change the <u>distribution</u> of its output, the higher the irradiance, the narrower the spread. We call the theory Intensity-Dependent Spread (IDS).

Results

Applying this theory as an image processing algorithm generates a number of surprizing and useful results.

First, the input is band-pass filtered, thus producing "edge enhancement". But unlike linear band-pass filtering eg. convolution with a difference of Gaussians, the peak frequency of the pass band shifts in direct proportion to the square root of the ambient irradiance. One important consequence is that, for quantum-limited detectors, the decrease in image S/N that occurs as scene irradiance is lowered is exactly balanced by an increase in the area over which signals are summed, maintaining a constant output S/N for all scene irradiances. Another consequence is that the resolution of the system is greater where the scene irradiance is greater.

Second, the amplitude of the response to an input edge depends on the ratio of the irradiances on the two sides of the edge, not on their difference or their absolute levels. Therefore, an IDS-processed image contains information about relative reflectances in the scene, and these reflectances can thus be recovered without regard for differences in scene irradiance from one time to another or from one region of the scene to another. This property, in turn, provides "brightness constancy", and when applied to each image separately in multispectral imagery, it results in color constancy.

Temporal Properties

In the preceding, it was implied that the SSD's are set up instantaneously, but adding the simple assumption that, instead, the lateral spread of signal from each input point occurs with a constant velocity, as it might if propagated by neurons, for example, then a large set of interesting temporal properties emerge. These properties are the temporal analogs of the spatial properties described above.

If the irradiance of a uniform input field is varied over time, the result is temporal band-pass filtered, giving on and off transients, for example. Further, the temporal frequency of the pass-band increases with the square root of the mean irradiance. This automatically increases the period of temporal summation at low light levels, and provides higher temporal resolution for higher scene irradiances.

Computation

The mathematics to describe IDS are only tractable for a few relatively uninteresting input patterns. Therefore, to use the technique on real images, we simulate the theory itself. That is fairly easy to do, but is slow on a standard computer. For example, processing a 512x512x8bit image of the USC girl on a Sun 3160 system requires about 40 minutes. However, working under a NASA grant, Odetics, Inc. of Anaheim, Ca., has developed a board set that, when used with Data Cube hardware, will process that image in about 4 seconds and they expect to have a board set in about a year that will do it in 1/30 sec.

Relation to Human Vision

Those familiar with human vision will recognize that the results of IDS describe a wide variety of phenomena known for human vision. Some of them are Mach Bands, the Craik-O'Brien illusion, band-pass spatial Contrast Sensitivity Functions that shift with illuminance, increases in acuity with illuminance, Ricco's Law, changes in Ricco's Law with illuminance, "early" light and dark adaptation, Bloch's Law, changes in Bloch's Law with illuminance, band-pass temporal Contrast Sensitivity Functions that shift with illuminance, changes in reaction time with illuminance, and many others.

Although we do not have time to describe these relationships in detail, we do want to point out that, in addition to its uses in image processing, IDS provides a concise model that describes a large majority of the known phenomena of human brightness perception and so it may also be useful in that way to people working in applied vision.

References

1) Cornsweet, T. N. "Prentice Award Lecture: A simple retinal mechanism that has complex and profound effects on perception", Amer. J. Optom. & Physiol. Optics, **62**, 427 (1985).

2) Cornsweet, T. N. and Yellott, J.I.Jr., "Intensity-Dependent Spatial Summation", J. Opt. Soc. Amer. A, **2**, 1769 (1985).

3) Yellott, J. I., Jr. "Photon Noise and Constant Volume Operators", J. Opt. Soc. Amer. A, **4**, 2418 (1987).

IMAGE CHARACTERISTICS RECOVERY
FROM BANDPASS FILTERING

Rachel Alter-Gartenberg[*]
Old Dominion University, Norfolk, Virginia 23529

Ramkumar Narayanswamy
Science and Technology Corporation, Hampton, Virginia 23666

ABSTRACT

Images filtered with bandpass filters preserve most of their original characteristics. Quantifying them, we recover the original intensity and reflectance representations from the bandpassed data.

1. INTRODUCTION

A local change in intensity (edge) is an image characteristic that is preserved when filtered through a bandpass filter. Primal sketch representation of the image, that results from this preservation property for low level vision and other applications, was discussed in many papers since Marr proposed his model for early human vision.[1] In this presentation we move beyond primal sketch extraction to the recovery of intensity and reflectance representations from the bandpass filtered image data.

The following isotropic bandpass filters are the usually accepted models proposed for the retinal processing in human vision: the spatially-invariant Laplacian of Gaussian ($\nabla^2 G$) filter[2] and the spatially-variant Intensity-Dependent Spatial Summation (IDS) filter.[3] In what follows, we focus on both these filters. The $\nabla^2 G$ filter preserves the change of the intensity at the edge location. Therefore, we can recover the (relative) intensity representation of the original image from the $\nabla^2 G$ bandpassed data alone.[4] The IDS filter preserves the actual intensities on both sides of the edge, and the reflectance ratio at the edge location. The recovery of the (relative) reflectance representation is of special interest as it erases shadowing degradations and other dependencies on the temporal illumination.[5] We delineate the condition under which these properties can be recovered and assess the quality of that recovery. In addition, we developed a coding scheme to transmit only the information associated with the edge location that is necessary for the recovery process. The potential for high data compression ratio and the recovery quality, as assessed by the correlation between the original target and the recovered one, offers a new approach to low level vision processing as well as to high-data compression coding applications.

2. THE LAPLACIAN OF GAUSSIAN FILTER

Lateral inhibition in early human vision processing was used by Marr and Hildreth[2] as a basis for developing the $\nabla^2 G$ filter:

$$\tau(x,y;\sigma) = \frac{1}{\pi\sigma^4}\left(1 - \frac{r^2}{2\sigma^2}\right)\exp\left(-\frac{r^2}{2\sigma^2}\right), \quad (1)$$

where $r^2 = x^2 + y^2$ and σ is the standard deviation of a normal distribution. The lateral inhibition model is approximated by convolving the image with the $\nabla^2 G$ operator. The resultant bandpassed data preserves the location of changes in the image intensity together with the magnitudes of those changes (ΔI).

Isotropic filters allow us to choose a one-dimensional change of intensity from I to $I + \Delta I$ to simulate an ideal edge. The response of such an image to the $\nabla^2 G$ filter, as given by

$$S(x;\sigma,\Delta I) = \frac{\Delta I \cdot x}{\sqrt{2\pi}\sigma^3}\exp\left(-\frac{x^2}{2\sigma^2}\right), \quad (2)$$

is a Mach band signal around the edge boundary that crosses zero exactly at the location of the change of the intensity. The corresponding peak and trough are symmetrically located at $x = \sigma$ and $x = -\sigma$, perpendicular to the local edge direction, taking the values

$$S(\pm\sigma) = \pm\frac{\Delta I}{\sqrt{2\pi e}\sigma^2}.$$

The Mach band amplitude $A = S(\sigma) - S(-\sigma)$ becomes ΔI for $\sigma_o = \left(\frac{2}{\pi e}\right)^{1/4} \cong 0.69$ which corresponds to Marr's smallest channel in early vision[6], when normalized relative to the unit sampling interval.[7]

To maintain a constant amplitude ΔI, regardless of the operator size (the choice of σ), the bandpassed image data should be multiplied by the normalization factor σ^2/σ_o^2. Figure 1 shows the normalized response of an edge to different operator sizes. The Mach band amplitude is exactly ΔI for all the responses and can be measured directly through the values of $S(x = \pm\sigma)$. By measuring the values $S(x = \pm\sigma)$ of the bandpassed data, we can recover the (relative) intensity representation of the original image.

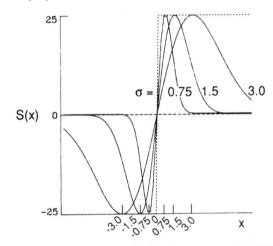

Figure 1: Normalized response of an ideal edge to three $\nabla^2 G$ filters.

The ΔI measurement accuracy depends on the discrete digital implementation of the continuous model and on the spatial distance between edges relative to the filter size. Spatial details smaller than 3σ are blurred by the filtering process and therefore cannot be retrieved by this recovery process. Details larger than 3σ are recoverable by measuring the peak and trough values of the blurred signal. The actual peak or trough location may fall in between two discrete processing intervals for some implementations or edge orientations. On such cases, the change of intensity is measured through the response of $S(x)$ as given by

[*]Mail address: NASA Langley Research Center, MS 473, Hampton, VA 23665.

$$\Delta I = \frac{2}{b} S(x) \exp\left(\frac{1-b^2}{2b^2}\right), \tag{3}$$

where $b = \sigma/x$ and $x = \sigma$ is the ideal case.

The relative error in measuring $S(x)$ is given by

$$\frac{\epsilon_s}{S(x)} = \left(1 - \frac{1}{b^2}\right)\epsilon_{\Delta x},$$

where $\epsilon_{\Delta x}$ is a function of the interval of processing Δx, i.e., the discretization interval transforming the continuous case into a digital discrete process. The corresponding relative error in measuring ΔI from the Mach-band response becomes

$$\frac{\epsilon_{\Delta I}}{\Delta I} = \left[\left(1 - \frac{1}{b^2}\right)^2 + \frac{1}{\sigma^2}\left(b + \frac{1}{b}\right)^2\right]^{1/2}\epsilon_{\Delta x}.$$

The error in estimating ΔI relies directly on $\epsilon_{\Delta x}/\sigma$. Therefore an accurate estimation of ΔI requires the interval of processing to be chosen such that the distance between the zero-crossing and peak locations will be at least two intervals of processing.[4,8]

3. THE INTENSITY-DEPENDENT SPATIAL SUMMATION FILTER

Adaptive response to the intrinsic noisiness of light (photon noise) in early human vision processing was used by Cornsweet and Yellott[3] as a basis for developing the IDS filter. Assuming that lateral inhibition summation is of adaptive nature, the IDS filter consists of nonnegative, spatially homogeneous, circularly symmetric point spread functions (PSF) K, with unity volumes. The PSF's differ from each other by their spreads which are inversely dependent on the local intensity $I(x,y)$ as given by

$$\tau(x, y; I) = I \, K(Ir^2), \tag{4}$$

where $r^2 = x^2 + y^2$. Thus, the effect of the input intensity is to rescale the PSF's, leaving their basic form unchanged. The image response to the IDS model is the sum of the PSF's. The resultant bandpassed data preserves not only the edge location but the original intensities that created the edge, I and $I + \Delta I$ and the corresponding Weber fraction (reflectance ratio) $W = \Delta I/I$.

The response of an ideal edge to the IDS filter,

$$S(x; I, \Delta I) = 1 - \int_0^{\sqrt{I}x} K(z)dz + \int_0^{\sqrt{I+\Delta I}x} K(z)dz, \tag{5}$$

is a Mach-band signal around the edge boundary that crosses the value one exactly at the edge location. The corresponding peak and trough are symmetrically located at a distance p that satisfies the equation $\sqrt{1+W} \cdot K\left(\sqrt{I+\Delta I}\, p\right) = K\left(\sqrt{I}p\right)$, and take on the value $S(p; I, \Delta I)$.

The following separable PSF meets the IDS model requirements and has a finite support which assures accurate discrete implementation of Eq. 5.

$$K(x,y) = \begin{cases} 1, & 0 \le \sqrt{x^2+y^2} \le 1/\sqrt{\pi} \\ 0, & \text{elsewhere} \end{cases}.$$

Solving Eq. 5 for the PSF defined above shows the Mach-band peak to be located at a distance $p = [\pi I(2+W)]^{-1/2}$ perpendicular to the local edge direction,[5] taking the value

$$S(p) = 1 + \frac{1}{\pi}\sin^{-1}\left(\frac{W+1}{W+2}\right)^{1/2} - \frac{1}{\pi}\sin^{-1}\left(\frac{1}{W+2}\right)^{1/2} \tag{6}$$

Consequently, Eq. 6 shows that the amplitude of the Mach-band is a function of W (the reflectance ratio) while its spread is also a function of the original intensity I, as illustrated in Figure 2.

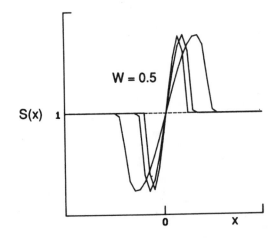

Figure 2: IDS response to edges with a constant Weber fraction W = 0.5.

In practice, measuring p and $S(p)$ enables us to recover the intensity and (relative) reflectance representations by estimating W, I and ΔI as:

$$W = \frac{2\sin\theta}{1-\sin\theta}, \quad I = \frac{1}{2p^2(W+2)}, \quad \Delta I = WI \tag{7}$$

where $\theta = \pi[S(p) - 1]$.

The original intensity can be recovered for spatial details that are larger than $2[\pi\Delta x(\Delta I + 2I)]^{-1/2}$ intervals of processing by measuring p and $S(p)$ around the edges, where Δx is the distance between the intervals of processing. The accuracy with which $S(p)$ is measured directly affects the estimate of the reflectance ratio:

$$\epsilon_W = \frac{2\pi\cos\theta}{(1-\sin\theta)^2}\epsilon_s.$$

The accuracy with which the intensities I and ΔI can be estimated is a function of the accuracy by which p and $S(p)$ are measured, and the corresponding W is estimated, as given by

$$\frac{\epsilon_I}{I} = \left[(1+W)\pi^2\epsilon_s^2 + 4\pi I(W+2)\epsilon_{\Delta p}^2\right]^{1/2}$$

and

$$\frac{\epsilon_{\Delta I}}{\Delta I} = 2\left[\frac{\pi^2(W+1)}{W^2}\epsilon_s^2 + \frac{\pi\Delta I(W+2)}{W}\epsilon_{\Delta p}^2\right]^{1/2}$$

where $\epsilon_{\Delta p}$ is $\pm 0.5\Delta x$ at most.

4. RESULTS

Figures 3 and 4 summarize the recovery of image characteristics from the bandpassed data. In Fig. 3 the targets are computer-generated targets while in Fig. 4 it comes from an experimental setup. The first stage in the recovery process is to extract the location of intensity transitions (edges) through the zero crossings of the $\nabla^2 G$ bandpassed signal or the one crossings of the IDS bandpassed signal.[8] The resulting primal sketches are illustrated in Fig. 3(b) and 4(b). The recovery of the (relative) intensities from the $\nabla^2 G$ bandpassed signals[4] are illustrated in Fig. 3(c) and 4(c). The recovery of the intensities and the (relative) reflectances from the IDS bandpassed signals[5] are illustrated in (d) and (e), respectively. The quality of the recovery is measured by the cross correlation factor ρ, between the original image and the recovered one.

The high correlation between the original target and the recovered one suggests that the recovery process may be used as a decoder for a coding scheme in which only the edge locations and the necessary associated information are transmitted.[9] In these two figures, the compression ratios for such a coding scheme are labeled C_D for the $\nabla^2 G$ primal sketch coding scheme and C_I for the IDS primal sketch coding scheme. These ratios should be compared to the Huffman[10] code of the original targets labeled by C_0. The compression ratios are higher by orders of magnitudes while the loss, as is denoted by ρ and by visual inspection, is relatively insignificant.

5. CONCLUDING REMARKS

Assessment of the response of an ideal edge to a bandpass filter reveals that most of the target's characteristics are preserved. For applications that use bandpassed signals, the recovery of those characteristics offers new dimensions to image understanding, requiring minimal extra calculations based on the existing bandpassed data information. Furthermore, these results may be helpful in high data compression coding applications, especially when high data rate transmission is required.

REFERENCES

1. D. Marr, S. Ullman and T. Poggio, "Bandpass channels, zero-crossings, and early vision information processing," J. Opt. Soc. Am., Vol. 69, No. 6, pp. 914-916, 1979.

2. D. Marr and E. Hildreth, "Theory of edge detection," Proc. R. Soc., London B 207, pp. 187-217, 1980.

3. T. N. Cornsweet and J. I. Yellott, Jr., "Intensity-dependent spatial summation," J. Opt. Soc. Am. 2, pp. 1769-1786, 1985.

4. R. Alter-Gartenberg and F. O. Huck, "Beyond primal sketches," Computer Vision, Graphics, and Image Processing, submitted, (1989).

5. R. Alter-Gartenberg and F. O. Huck, "Feature extraction with intensity-dependent spatial summation," J. Opt. Soc. Am., submitted, (1988).

6. D. Marr, T. A. Poggio, and E. Hildreth, "Smallest channel in early human vision," J. Opt. Soc. Am., Vol. 70, No. 7, pp. 868-870, July 1980.

7. F. O. Huck, C. L. Fales, D. J. Jobson, S. K. Park, and R. W. Samms, "Image-plane processing of visual information," Applied Optics, Vol. 23, No. 18, pp. 3160-3167, 1984.

8. R. Alter-Gartenberg, C. L. Fales, F. O. Huck, and J. A. McCormick, "End-to-end performance of image gathering and processing for edge detection," Computer Vision, Graphics, and Image Processing, submitted, (1987).

9. R. Narayanswamy, R. Alter-Gartenberg, "Coding of images using edge associated information," (in preparation).

10. D. A. Huffman, "A method for the construction of minimum redundancy codes," Proc. IRE 40, no. 9, 1098-1101, 1952.

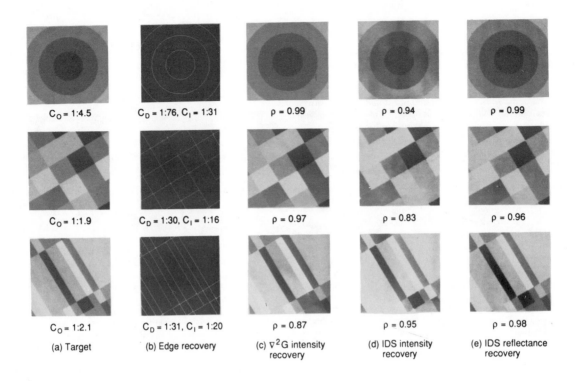

$C_O = 1{:}4.5$ $C_D = 1{:}76, C_I = 1{:}31$ $\rho = 0.99$ $\rho = 0.94$ $\rho = 0.99$

$C_O = 1{:}1.9$ $C_D = 1{:}30, C_I = 1{:}16$ $\rho = 0.97$ $\rho = 0.83$ $\rho = 0.96$

$C_O = 1{:}2.1$ $C_D = 1{:}31, C_I = 1{:}20$ $\rho = 0.87$ $\rho = 0.95$ $\rho = 0.98$

(a) Target   (b) Edge recovery   (c) $\nabla^2 G$ intensity recovery   (d) IDS intensity recovery   (e) IDS reflectance recovery

Figure 3: Recovery of computer generated targets.

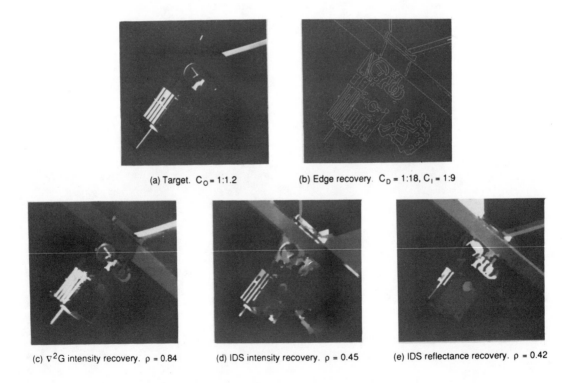

(a) Target. $C_O = 1{:}1.2$ (b) Edge recovery. $C_D = 1{:}18, C_I = 1{:}9$

(c) $\nabla^2 G$ intensity recovery. $\rho = 0.84$ (d) IDS intensity recovery. $\rho = 0.45$ (e) IDS reflectance recovery. $\rho = 0.42$

Figure 4: Image recovery from the bandpass filtered data.

Intelligent Temporal Subsampling of American Sign Language Using Event Boundaries

David H. Parish[1], George Sperling, and Michael S. Landy
New York University, 6 Washington Place, New York, N.Y., 10003

Introduction

American Sign Language (ASL) is a gestural form of communication used by the North American deaf and hearing impaired communities. In free conversation, ASL is as rapid a form of communication as most spoken languages, including English. Yet, users of ASL are prevented from using their most efficient form of communication over long distances by the absence of affordable, high bandwidth communication technology. Use of the existing, low bandwidth telephone switching network for ASL video transmission requires substantial signal compression. To insure that the ASL signal remains useful following compression, we must first determine the visual requirements for intelligible ASL.

Previous efforts to determine the minimal ASL representation have focussed primarily on spatial image compression schemes, though temporal subsampling was occassionally investigated (Pearson, 1981). Sperling, Landy, Cohen and Pavel (1985) measured ASL intelligibility as a function of spatial bandwidth, demonstrating that ASL remained highly intelligible despite very severe compression distortions. In the present paper, we address the issue of temporal compression. By identifying redundant dynamic information, we try to intelligently subsample the ASL signal in time, thereby retaining intelligibility while reducing the number of to-be-transmitted frames.

Temporal Compression

We consider an ASL sequence as a series of events that unfold over time. The distinction between a *motion* and an *event* is important; an event is a sequence of motions that fall under a common label. The question of temporal compression reduces to one of understanding how events are most efficiently represented, a problem that has received considerable attention in the field of social psychology. It has been suggested that event boundaries are the critical components in conveying the global activity within a sequence (Newtson, 1973). A series of experiments by Newtson and his many collaborators demonstrate the perceptual salience and importance of such boundaries.[2]

To test the hypothesis that event boundaries are more informative than non-boundaries, and therefore form a basis for temporal subsampling, we measured intelligibility for temporally subsampled ASL sequences. Subampling was based upon either the location of event boundaries within the sequence or upon constant subsampling, a scheme that simply chose every m^{th} frame. Event boundaries were located by noting that they tended to occur at starts, stops and changes in direction of the moving object(s). A simple algorithm, called the activity-index, was used to analyze each sequence, locate the event boundaries, and select the corresponding frames.

The Activity-Index. The stimulus set consisted of 84 isolated ASL signs that had previously been filmed and digitized by Sperling et al. (1985); each sign was a sequence of 60 frames. A single signer was filmed from the waist up under conditions in which the hands and face appeared in high contrast to the surround and body. For a sequence that consists of N frames of X rows by Y columns, the

---

[1]current address: U. of Minnesota, 75 East River Rd., Mpls., MN. 55455

[2]See Newtson, Hairfield, Bloomingdale and Cutino (1987) for a review.

luminance value at spatial location x,y in frame n, $1 \leq n \leq N$, is $l(x,y,n)$. The activity-index simply counts the numbers of pixels in each frame that undergo a luminance change greater than θ, by setting

$$t_\theta(x,y,n) \;=\; \begin{cases} 1 & \text{if } |l(x,y,n) - l(x,y,n-1)| > \theta \\ 0 & \text{otherwise.} \end{cases}$$

The activity index $a_\theta(n)$ is the fraction of pixels in frame n that experience a suprathreshold change in luminance:

$$a_\theta(n) \;=\; \frac{\displaystyle\sum_{x=1}^{X}\sum_{y=1}^{Y} t_\theta(x,y,n)}{XY} \;.$$

The higher the threshold parameter θ, the smaller the influence of pixels that change as a result of camera or digitizing noise rather than as the result of a moving object. Starts, stops and changes in direction correspond to local minima when the activity-index is plotted as a function of frame number; event boundaries were located by finding the local minima in the activity-index function. A second parameter, α, was used to control the coarseness of sampling by the activity index. α specifies the minimum increase in $a_\theta(n)$ that must occur between consecutively chosen frames.

Constant Subsampling. Constant subsampling, in which every m^{th} frame is chosen, $2 \leq m \leq N$, was used as a control technique. This method is arbitrary in the sense that it is completely dependent upon the selection of m and the starting point within the sequence of frames. If we assume that certain frames within a sequence are more informative than others, then we would expect constant subsampling to do poorly, that is, to produce unintelligible sequences, as, for any m, all frames have an equal probability of being discarded.

Frame Re-scaling. Having selected a subset of frames from a sequence, it is desirable to scale the duration of each sampled frame in time so that the reconstructed sequence retains as much as the temporal and rhythmic properties of the original sequence as possible. For constant subsampling, the number of repetitions for each frame was set equal to m, the constant sampling factor. For activity-index subsampling, frame repetition varied according to the location of the chosen frame in the original sequence.

Experiment 1

Procedure

The intelligibility of isolated ASL signs was measured. There were two subsampling conditions, activity-index and constant, and four compression factors within each subsampling condition (i.e., four values of α for the activity-index and four values of m for constant subsampling). We refer to the different compression factors as a variation in *frame-rate*, measured in frames per second. In actuality, all sequences were presented at 60 frame/sec. In the context of this study, frame-rate refers to the number of *new* frames displayed each second; frame repetition requires negligible overhead. Over the course of the experiment, all signs appeared in all conditions, although each subject saw each sign only one time. Eight ASL-fluent subjects participated in the study. All signs were stored and processed on a Vax 750 computer with an Adage RDS-3000 image processing system. For all conditions, subjects were required to respond to each ASL presentation with an English interpretation.

Results

The results of Experiment 1 are plotted in Figure 1, which shows the probability of correctly identifying an isolated ASL sign as a function of frame-rate (averaged over all signs) for the two sampling conditions. Except for the lowest sampling rate, activity-index subsampling produced sequences that were significantly more intelligible than those produced by constant subsampling. The cross-over interaction at the lowest sampling rate is attributed to a correctable sampling artifact of the activity-index scheme. Over the remaining portion of the curves, the activity-index scheme is approximately 15% better than constant subsampling, a highly significant improvement in the context of a communication system.

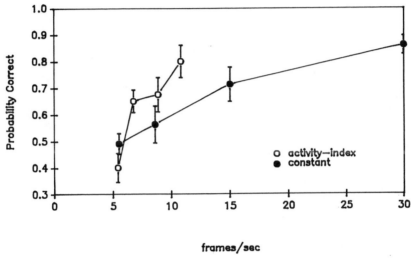

Figure 1. Probability of correctly identifying an isolated ASL sign as a function of frame-rate for activity-index and constant subsampling.

Experiment 2

There are compelling reasons for studying the intelligibility of ASL when presented in static, rather than dynamic, form. Most importantly, static images are used in most, if not all, ASL textbooks and dictionaries. Is the optimal subset of frames for dynamic display also optimal for static display? To answer this question, a second experiment was conducted in which a new group of 8 ASL-fluent subjects were required to interpret subsampled ASL signs presented statically. The frames of each sequence were arranged on a "page" in rows and columns, from left to right and from top to bottom of the page, much like the frames of a comic-strip. The stimuli and conditions were exactly those used in Experiment 1. Subjects were again required to interpret each "page" as it appeared. There was no time constraint on the subjects' responses.

Results

The results of the second experiment are plotted in Figure 2 which, again, shows the probability of correctly identifying a sign as a function of frame-rate for the two sampling conditions. Here we see that activity-index performance rises above that of constant subsampled images as the frame-rate *decreases* down to 6.75 frames/second. At peak intelligibility, activity-index performance is an 20% better than that of constant subsampling. It is interesting to note that for static presentation, over a limited range of frame-rates, the presence of unnecessary frames actually degrades performance.

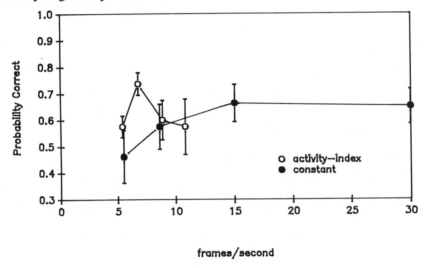

Figure 2. Probability of correctly identifying an isolated ASL sign as a function of frame-rate for activity-index and constant subsampling for static images.

Conclusions

Temporal subsampling that attends to the activity within the sequence produces compressed sequences that are significantly more intelligible at low frame-rates than arbitrary temporal subsampling. Specifically, the use of event boundaries as a basis for temporal segmentation offers a model within which segmentation can occur. The activity-index, though extremely simple, worked well in locating event boundaries in the highly restricted domain of the present study. A more sophisticated model would have to be developed to handle a "real-world" environment.

References

Newtson, D. (1973). Attribution and the unit of perception of ongoing behavior. *Journal of Personality and Social Psychology, 28,*28-38.

Newtson, D., Hairfield, J., Bloomingdale, J., & Cutino, S. (1987). The structure of action and interaction. *Social Cognition, 5,* 191-237.

Pearson, D.E. (1981). Visual communication systems for the deaf. *IEEE Transactions on Communications, 29,* 1986-1992.

Sperling, G., Landy, M.S., Cohen, Y., & Pavel, M. (1985). Intelligible encoding of ASL image sequences at extremely low information rates. *Computer Vision, Graphics, and Image Processing, 31,*335-391.

Image Compression in Noise

Scott Daly

Hybrid Imaging Systems Division - Photographic Research Laboratories
Eastman Kodak Company, Rochester, New York 14650

Use of the CSF in Image Compression

The visual system's variations in sensitivity to spatial frequencies are critical to any imaging system where the image is to be displayed and viewed by a human observer. These variations are described by the contrast sensitivity function (CSF) which has found wide application in image compression schemes using the discrete cosine transform (DCT) [1-3], vector quantization [4], and spatial filter hierarchies [5]. All of these approaches provide access to the frequency domain, and the use of the CSF becomes straightforward in controlling the quantization process of the algorithm. Quantization is used to code the algorithm coefficients or vectors; an increase in the size of the quantization interval reduces the entropy, and thus the bit rate. However, larger quantization intervals increase the quantization error of the algorithm, which can be regarded as noise that will degrade the image if it is visible. This quantization noise must be detected in the presence of the effective internal noise of the visual system, which is proportional to the inverse of the CSF. Therefore, the inverse CSF can be used to scale the quantization intervals, allowing larger intervals for frequencies where the visual system is less sensitive and smaller intervals where it is more sensitive. If, for all frequencies, the maximum error of the frequency specific quantization noise is kept less than the effective internal noise of the frequency, the compressed image will be visually indistinguishable from the uncompressed image. We refer to this condition as *perceptually lossless*, as opposed to *mathematically lossless*, in which the digital code values are exactly preserved. Since the bit rates for perceptually lossless compression are less than a quarter of those for mathematically lossless compression, perceptually lossless compression is a useful criterion when the image is to be viewed by human observers.

Noise Adaptive Model of the Visual CSF

The CSF is highly adaptive, depending on such parameters as light adaptation, color, and noise. This work will describe the effects due to broadband (i.e, white) 1-D and 2-D noise, while the other adaptive parameters will be held constant. Commonly, the CSF is measured in uniform fields free of noise (except for the quantum fluctuations of light). However, as the background noise level is increased, the CSF changes in both shape and peak sensitivity, and these changes are important to image compression. Several terms need to be introduced in order to quantify noise. The first of these is rms contrast, C_{RMS}:

$$C_{RMS} = \sigma/\overline{L} \tag{1}$$

where σ is the standard deviation of a waveform when biased about its mean, $\overline{L}$, both in luminance units. With this contrast definition, arbitrary waveforms including noise can be described by their standard deviations in contrast. The rms contrast is related to another term, spectral density, $n^2(u)$, which describes the noise power at a specific frequency, u:

$$C_{RMS} = \sqrt{\int_{-u_{max}}^{+u_{max}} n^2(u)du} \tag{2}$$

Although rms contrast, C_{RMS}, is the most often used noise descriptor, the more frequency-specific spectral density has advantages due to the cut-off frequency of the CSF. To retain some familiarity with rms contrast, we will take the square root of spectral density, which is linearly related to rms contrast per spatial frequency. Since we are only addressing white noise (up to a cut-off frequency), the values for $n_{RMS}(u)$ will be the same for all frequencies, and we will refer to it as simply n_{RMS}.

To understand the changes in the CSF due to noise, we can examine the effect of the magnitude of broadband noise on the normalized sensitivity of a single frequency (4 cpd), shown as the solid line in Figure 1. Note the two dotted asymptotic regions in this log-log plot. For low noise levels, the data approach a constant sensitivity that is equal to the sensitivity of the frequency in a uniform field. For high noise levels, the data obey Weber law behavior, approaching an asymptote with a slope of -1.0 in this plot format. Studies which agree with the shape of this plot include 1-D dynamic noise [6,7], 1-D static noise [8], and 2-D static noise [9]. These changes in sensitivity as a function of noise level for a single frequency can be modelled as

$$S(n_{RMS}) = \left(W_N\sqrt{n_{CRIT}^2 + n_{RMS}^2}\right)^{-1} \tag{3}$$

where $S(n_{RMS})$ is the sensitivity as a function of broadband noise, n_{RMS}, while the critical noise, n_{CRIT}, is the noise level corresponding to the intersection of the two asymptotes. The noise Weber fraction, W_N, describes the vertical location of the Weber law asymptote and can be found from data when the background noise is much greater than the critical noise:

$$W_N = (n_{RMS} \cdot S(n_{RMS}))^{-1} \quad \text{for} \quad n_{RMS} \gg n_{CRIT} \tag{4}$$

This overall relationship also occurs for other spatial frequencies, shown dashed in Figure 1. Potentially, both the critical noise, n_{CRIT}, and the Weber noise fraction, W_N, can be functions of spatial frequency in this model. However, in some of the available data [6,8] the Weber noise fraction is invariant with frequency, while in others [7,9] it increases with spatial frequency. Due to a lack of sufficient data, we decided to assume that the noise Weber fraction is invariant, and as a result the data converge for high noise values as shown in Figure 1. Sensitivity as a function of both frequency and background noise can then be modelled as

$$S(u, n_{RMS}) = \left(W_N\sqrt{n_{CRIT}^2(u) + n_{RMS}^2}\right)^{-1} \tag{5}$$

The functional dependence of the critical noise on frequency can be easily determined by solving Eq. (5) with the background noise, n_{RMS}, equal to zero:

$$n_{CRIT}(u) = (W_N \cdot S(u, 0))^{-1} \tag{6}$$

where $S(u, 0)$ is the CSF in a uniform field (zero noise contrast).

Using a Weber noise fraction determined from a study using 1-D static white noise [8] and a model for the noise free CSF, $S(u, 0)$, to determine the critical noise, n_{CRIT}, the model of Eq. (5) was used to calculate sensitivity as a function of frequency for a series of noise levels, shown in Figure 2. As the background noise level is increased, the CSF undergoes a reduction in sensitivity and a flattening in shape such that there is less difference in sensitivity across the frequencies. These results are similar to the more common studies of suprathreshold sensitivity [10].

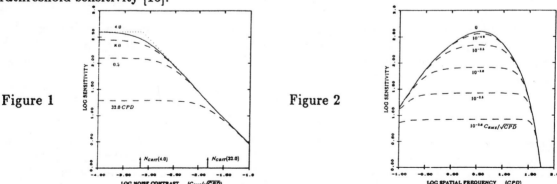

Figure 1   Figure 2

Application of Noise Adaptive CSF to Image Data Compression

There are two main sources of noise in digital imagery relevant to image compression. The first of these is noise introduced into the image prior to the compression process and may be termed source noise. The second and most relevant noise source is the noise introduced at the display stage, as the viewed image will always be seen through the display noise. If these noise sources are white, they can be used as the background noise level for calculating the noise adaptive CSF. To apply the noise adaptive CSF to a compression algorithm, the contrast per digital code value relationship of the display needs to be measured. Then, for a given compression algorithm, there will be a known relationship between the quantization of a coefficient or vector and the resulting contrast of its basis function or template, which will be limited in spatial frequency. The CSF is then used to determine the maximum allowable error in the contrast of this basis function such that the error is below visual threshold.

The model of the noise adaptive CSF was tested in an image compression algorithm that utilized the DCT. The particular implementation is similar to that of the currently proposed CCITT image compression standard [3], with a few minor exceptions. Nine noise levels were tested, and for the main experiment a CSF array calculated for each noise level was used in the compression. Input images of different noise levels were modelled by digitally adding 2-D white gaussian noise to a relatively noise-free original image, while display systems with different noise levels were modelled by digitally adding similar noise fields to the image prior to display. In Figure 3, the block diagram describes the experiments. The quantization of the compression process is controlled by the CSF model, which in turn was controlled by the noise level of the input or display noise model. Images compressed using the noise adaptive CSF model were compared against images using a CSF model with no noise adaptivity, and bit rates and image quality were compared. For the input noise case, the noise adaptive CSF resulted in bit rates that remained fairly constant despite the increases in image noise, while the bit rates for the non-adaptive model increased significantly, by more than a factor of 3 for the highest noise level studied. These results for one image are indicated in Table 1. When compared to the original in the same noise, it was difficult to discriminate between the compressed and the original, though we leave the perceptual results to the observer study in the next section. For the display noise study, the sensitivity decreases of the noise adaptive CSF allow for more quantization error as the noise level increases, which allows the bit rates to be significantly reduced while the bit rate for the non-adaptive scheme now remains unchanged.

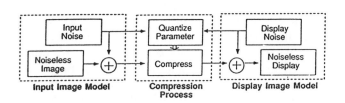

IMAGE: balloons					
NOISE LEVELS		BIT RATES (BPP)			
N_{rms}	S.D.	INPUT NOISE		DISPLAY NOISE	
log	code value	adaptive	nonadaptive	adaptive	nonadaptive
$-\infty$	0.0	1.02	1.02	1.02	1.02
-4.05	0.289	1.02	1.02	1.00	1.02
-3.50	1.0	1.03	1.07	0.98	1.02
-3.25	1.8	1.04	1.17	0.90	1.02
-3.00	3.2	1.12	1.45	0.75	1.02
-2.75	5.7	1.26	1.90	0.57	1.02
-2.50	10.0	1.39	2.47	0.39	1.02
-2.25	18.0	1.51	3.20	0.24	1.02
-2.00	32.0	1.57	4.00	0.14	1.02

Figure 3   Observer Study   Table 1

Observer Study

The sensitivity curve in Figure 1 divides the noise-sensitivity plane into two main regions with the region below the curve corresponding to signals that are seen while those above the curve are unseen. The compressed images described in the previous section, in which a CSF corresponding to the noise level was used to control the quantization error, can be regarded as points falling along this curve where this curve now represents the CSF peak frequency sensitivity. In an observer study designed to test the model for display noise, 30 compressed images were created to span this plane, such that 8 images corresponded to points on this curve, while 14 were below and 8 were above the curve. Images residing below the curve correspond to images that were compressed with CSFs corresponding to an expected noise level but displayed with less noise, such that the point corresponding to the image is shifted to the left from the curve. In this case, the image error due to the compression process is expected to be more visible. Images with points above or to the right of the line correspond to adding more noise than used to calculate the CSF utilized in compression. In these images, the error due to the compression process is expected to be even less detectable. All points with identical vertical ordinates indicate the use of the same CSF. The noise level corresponding to that CSF is the abscissa of the intersection of the vertical ordinate with the modelled curve.

In the study, 42 observers were asked to make a 2AFC comparison between two images. They chose which of two images looked most like a third reference image. One of the tested images was the compressed image with a noise field added, such that it was a point in the plane described above. The other was the original image with the same noise field added. The reference image was another copy of the original with added noise. Thus, in 30 sets of three images, the observer task is to discriminate between the original and the compressed for a particular level of display noise. A contour plot showing the results appears in Figure 4. The values describe the probability of detection from 0.0 to 1.0, after correction for guessing, while the horizontal axis indicates the noise level tested and the vertical axis indicates the peak sensitivity of the CSF used in compression. The dotted line in this plot corresponds to the model for static 2-D noise, and below the line detectability increases while above the line it decreases, as expected from the model. Though the data is noisy due to limited experimental runs, it essentially supports the noise adaptive CSF model.

Figure 4

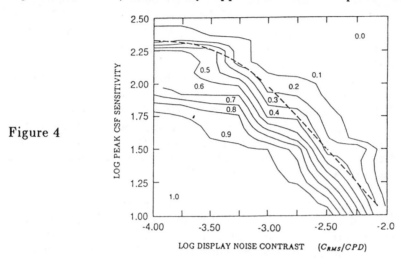

Conclusions

A model of the visual CSF has been developed that includes the effect of broadband noise on its shape and sensitivity. The model was developed from available data for the changes in sensitivity as a function of noise level for single spatial frequencies, and matches data for published measurements of the CSF in broadband noise. A simulation and observer study indicated that it is possible to extend this model to natural imagery for use in image compression. As the noise level increases, the advantage to using a noise-adaptive CSF model increases. However, the model is also significantly useful for extremely low noise levels since the use of a carefully calibrated noise-adaptive CSF can assure the designer that the optimal algorithm parameters are selected.

References

1. N. C. Griswold, Opt. Eng. V. 19 306-311 (1980).

2. N. B. Nill, IEEE V. COM-33 551-557 (1985) .

3. WG8, 'Initial draft for adaptive discrete cosine transform technique for still picture data compression standard', ISO/JTC1/SC2/WG8 N800 JPEG-5 May 1988.

4. K. S. Thyagarajan et al., IEEE ICASSP Proc. 141-144 (1985).

5. A. B. Watson, J. Opt. Soc. Am. A V. 4 2401-2417 (1987).

6. C. F. Stromeyer and B. Julesz, J. Opt. Soc. Am. V. 62 1221-1232 (1972).

7. D. G. Pelli, Ph.D. Dissertation, Cambridge University, UK (1981).

8. J. P. Thomas, J. Opt. Soc. Am. A V. 2 1586-1592 (1985).

9. S. Daly, SPIE Proc. 1077 in press.

10. M. A. Georgeson and G. D. Sullivan, J. Physiol. V. 252 627-656 (1975).

THURSDAY, JULY 13, 1989

8:30 AM–12:00 M

ThA1–ThA9

IMAGE QUALITY METRICS

Carlo Infante, Digital Equipment Corporation,
Presider

Effective Range of Viewing Instruments

Aart van Meeteren
TNO Institute for Perception
Kampweg 5, 3769 DE Soesterberg, The Netherlands

The distance at which an observer can just perform his particular visual observation task is perhaps the most practical metric of image quality with regard to viewing instruments. The direct relation between the distance of a target and the scaling of its image upon the retina suggests a simple theoretical approach: is it possible to express image quality in some sort of effective retinal "pixel-size", such as visual acuity, but then in a more generalized form, including the effects of luminance level, contrast, and noise? In this paper experiments will be described in order to determine the effective range first of a typical image intensifier system, and second for a thermal viewing system. The results will be discussed in the light of the above question.

In order to define a more or less realistic observation task for image intensifier systems a set of six different military vehicles in side view was specified, allowing for a number of discriminations with different levels of difficulty (for more detail see Van Meeteren, 1977). The principle of the experiments was to determine the percentage of correct identifications by repeated presentation of the objects of this set in random order. Obviously the percentage of correct identifications will be higher when more difficult discriminations can be made.

Vision through image intensifiers depends as much on the luminance level and on atmospheric contrast reduction as unaided vision. So, for a complete exploration it was decided to determine recognition distances as a function of luminance with contrast as a parameter. It is virtually impossible to conduct experiments of this nature and this extent directly in the field. The field-situation was therefore simulated by indoor slide-projection. Atmospheric contrast reduction was simulated by super-position of a uniform veiling light over the scenery. The experiments were arranged in sessions of 80 slide-presentations all at the same luminance level and with the same veiling light, but with three different distances. In this way the recognition distance corresponding with 50% correct identifications could be determined by interpolation for each session. Fig. 1 summarizes the resuslts. Note that each data point is based on 80 presentations.

The results are in good overall agreement with fundamental expectations. Vision through image intensifiers basically is photon-noise limited. As Fig. 1 illustrates recognition distance as well as contrast sensitivity both roughly increase with the square root of luminance. Above 2000 m the recognition distance begins to level off due to the optical resolution limit of the device.

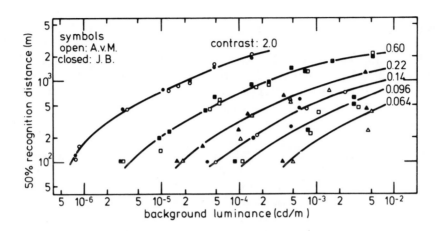

Fig. 1 Recognition distance for an image intensifier system

The results are replotted in Fig.2 in the form of contrast sensitivity against recognition distance with luminance as parameter. The curves represent independent additional measurements of the threshold-contrast for a circular disc upon a uniform background as a function of its diameter. Fig. 2 illustrates the best fit between these curves and the results of the recognition experiments, that could be obtained by suitable shifts along the axes. The fit suggests that recognition of the set of vehicles used in our experiments is comparable with the detection of a 0.7 m diameter circular disk. This is a very satisfactory result as the retinal image of such a disc can be considered exactly as the effective retinal "pixel-size" mentioned above.

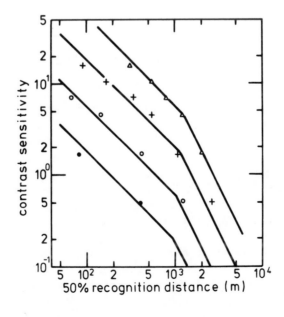

Fig.2 Data of Fig.1 replotted curves refer to the detection of a circular disc of 0.7 m.

The images produced by thermal viewing systems are in no way comparable with those coming from image intensifiers. Thermal radiation is
emitted by the scenery, whereas the light used by image intensifiers
is reflected light. Furhter, thermal viewing systems are scan-line
devices like television. Will it be spossible and practical to evaluate the range of thermal viewing systems in the same way as described
above for image intensifiers?

To investigate this, again a set of six military vehicles was thermo-
graphed in the field in front and side views and in two different
states of warming up. The primary effect of larger distance is a
lower scanline density measured in lines per meter on the target.
Fig. 3 illustrates how target identification improves with higher
scan-line density. Hot front views apparently cannot be completely
distinguished by their own nature. This certainly is a drawback of
thermal viewing systems. The potential use of these devices is more
clearly disclosed by the subset of warm side views.

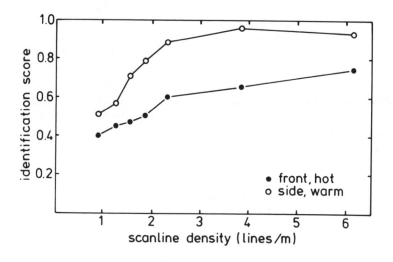

Fig.3 Percentage of correct identifications of thermal images
as a function of scanline-density

With a simple geometric conversion scan-line density can be replaced
by observation distance, but it must be realized that the image still
can be displayed with different angular magnifications. As to this we
have verified, that the only effect of angular magnification is the
trivial one: it must be large enough to avoid visual resolution problems. Thus, the range of a thermal viewing system, corresponding for
instance with 70% correct identifications of warm side views, can be
derived directly from Fig.3 as soon as the conversion factor given by
the design is known.

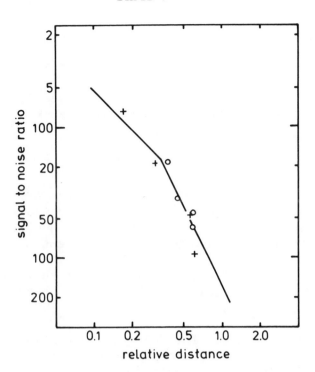

Fig.4 Relation between recognition distance and signal-to-noise ratio for a thermal viewing system. Curve refer to the detection of a circular disc of 0.3 m.

Although scan-line density is the most dominant limiting factor (Van Meeteren and Mangoubi, 1981), noise and low thermal contrast certainly can play an important secondary role, as Fig. 4 illustrates. On the analogy of the critical dimension for recognition with image intensifers (see Fig. 2) we have tried to fit the data again with the - in this case theoretical - curve for the detection of a circular disc, and the result suggests that 70% correct identification of the set of warm side views roughly corresponds with the detection of a 0.3 m disc.

Summarizing it appears that image quality might quite well be expressed in an effective retinal pixel element, representing resolution, low contrast, low luminance, and noise effects all together. This approach differs in a number of aspects from currently used models in applied vision, as will be discussed in the paper.

References

Van Meeteren, A., "Prediction of realistic visual tasks from image quality data", SPIE, Vol. 98, 58-64 (1977).

Van Meeteren, A.,and Mangoubi, S., "Recognition of thermal images: effects of scanline density and signal-to-noise ratio", SPIE, VOL 274, 230-238 (1981)

58

Perceptual Image Quality Metrics

Jacques A. J. Roufs, Huib de Ridder, Joyce Westerink
Institute voor Perceptie Onderzoek
Postbus 513
5600 MD Eindhoven
The Netherlands

There is a rapidly growing awareness that the quality experience of the human observer is the ultimate criterion for the technical quality of display systems, coding algorithms and image-processing systems.

In order to establish useful relations between perceptual quality and the physical image parameters, quantitative measures of subjective quality and its underlying dimensions are necessary.

In analyzing the different methods for doing so, a distinction between appreciation-oriented and performance-oriented quality problems was found to be useful. Examples of different methods for scaling of perceptual quality and the strength of its underlying factors, matching of iso-quality curves, scaling of impairment and also thresholds as the upper-limits of unwanted and lower-limits of wanted phenomena will be dealt with critically.

Quality Measures in Digital Halftones

by
Paul G. Roetling
Xerox Corporation
Webster Research Center
Bldg. W128-27E
800 Phillips Road
Webster, NY   14580

Introduction
Over about the last two decades the use of digitally generated halftone images has increased significantly. Halftone images which are created digitally from sampled and quantized images have some unique characteristics which affect the perceived image quality. Among these are the number of gray levels which can be reproduced at any spatial frequency, how the screen pattern chosen affects those gray levels, how tone reproduction is controlled and how the structured noise affects appearance. This paper provides an overview of some of the work on these effects.

The paper considers the visual system MTF as it relates to binary images. It then briefly reviews the digital halftone process and how the process affects gray level reproduction. Rendition of image detail is divided into several areas such that the properties of the halftone structure which are important in each area can be discussed, including the effects of the structured noise in the image. Tone reproduction is considered briefly.

Vision
The visual MTF has been measured by a number of workers, including Cornsweet [1]. Dooley [2] described a mathematical function which was an acceptable fit to Cornsweet's data. In a previous paper, this author [3] used their results to determine the number of gray levels which can be seen at any spatial frequency. This result, shown in Figure 1, was derived in a way that makes it an overestimate of the probable actual performance of the eye, but it nevertheless gives the general form well enough to relate to binary image quality.

It is useful to note that the frequency where only 2 levels can be seen, that is where a black and white input can just be seen, is commonly called the resolution limit. If we consider digital images, we need to sample at least two times the maximum resolved frequency. In reference 3, the volume under the levels vs. frequency curve is evaluated, showing that the visual system can see less than 3 bits/pixel on the average (for sampling at twice the resolution limit) and unpublished work using JNDs shows that the number is closer to 2 bits/pixel [4]. Thus, although digital images are often stored at 8 bits/pixel to capture enough gray levels at low spatial frequencies, image compression should be able to reduce total storage considerably without loss of visible quality. Digital halftones are sometimes considered a compressed form, as discussed next.

Halftone Process
The halftone process is basically a method for converting a gray value (eg. reflectance) into a probability of a binary pixel being white. Thus, over any area

59

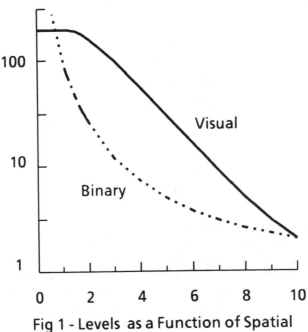

Fig 1 - Levels as a Function of Spatial
Frequency

consisting of several pixels, the average gray value of the binary image will match the original gray level of the multi-valued (here loosely called continuous-tone) image. Statistically, the larger the area one can average over, the closer the gray levels can be matched and therefore more different gray levels can be distinguished in lower spatial frequency components of a binary image. Again from reference 3, the dotted line in Figure 1 shows the limit of how many gray levels can be distinguished as a function of spatial frequency for an image sampled at twice the resolution limit of the eye.

A halftone image is created from the continuous-tone by adding a noise pattern and then thresholding the result to a binary image. The histogram of the noise controls the tone reproduction and the spatial statistics influence the detail in the image. Initially, in digital halftones, the noise pattern was chosen to be uniformly distributed white noise. The most common printed halftones now use a spatially repetitive pattern in which values are chosen to "grow" dots, that is, as the image darkens, the pixels that are black connect together into dots consisting of a number of adjacent pixels. The dots repeat at the spatial repeat pattern of the noise, usually 5 to 10 pixels or so (called n, below).

Image Rendition
Rendition of gray levels and detail in the halftone can be conveniently divided into 3 major areas, as shown in Figure 2 for a typical repetitive screen (noise pattern). Up to a spatial frequency of one-half the screen frequency, the halftone can represent $n^2 + 1$ gray levels (where n is the repeat length of the screen in pixels). One can easily understand this part of the space by thinking of each halftone dot as one sample of the original image, containing n^2 pixels which can be black or white. The quality tradeoff is the number of gray levels versus the frequency of the halftone screen pattern.

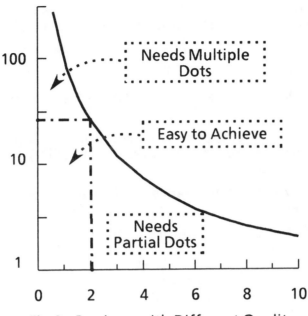

Fig 2 - Regions with Different Quality
Characteristics

At the higher spatial frequencies, gray levels change substantially in distances less than the size of each halftone dot. In this frequency region, detail is representable by the fact that whether a pixel is light or dark is a decision made pixel-by-pixel by a combination of screen and image values. For any screen with a dynamic range less than that of the continuous-tone image (the usual case), any pure black or white original pixel is correctly represented at that pixel. Thus, high contrast resolution matches the original. For intermediate gray values, how quickly a change of level causes a change in output depends on the screen pattern. For a single gray step change, the halftone will only be affected at one point per cycle, so the low contrast resolution falls to one half the screen frequency.

Finally, for spatial frequencies below one half the screen frequency, it is possible to represent more than the $n^2 + 1$ gray levels by introducing lower spatial frequency components into the screen pattern. This amounts to adding random or patterned noise, or in effect enlarging the repeat length of the screen, while leaving the majority of the spatial frequency power at the previous screen frequency. Here, the quality tradeoff is clearly the addition of more gray levels versus the perception of noise in the image.

Tone Reproduction
Most printers do not print spots which correspond precisely to the spacing between pixels. Thus, the theoretical tone reproduction assumed above does not usually occur. However, one can readily modify the histogram of the values in the screen (noise) pattern to compensate for the printer characteristics, so long as they are reproducible over time. Thus, the tone reproduction can be matched (at least at each achievable gray level) to the original continuous-tone image. Thus, quality related to tone reproduction resembles that for continuous-tone, except for the effects of the limited number of gray levels, as already discussed.

Conclusion

There are many more quality variables in halftones than discussed here. However, the intent was to show some of the unique problems of digital halftone image quality.

References

1. Cornsweet, T.N.: **Visual Perception**, Academic Press, New York, New York, 1971, p.330-342.

2. Dooley, R.P.: "Predicting Brightness Appearance at Edges Using Linear and Non-Linear Visual Describing Functions," presented at SPSE Annual Meeting, May 14, 1975, Denver, Colorado.

3. Roetling, P.G.: "Visual Performance and Image Coding," SPIE/OSA Vol.74 - Image Processing (1976)

4. Hamerly, J.: Private Communication

EVALUATION OF SUBJECTIVE IMAGE QUALITY WITH THE SQUARE ROOT INTEGRAL METHOD

Peter G.J. Barten

Barten Consultancy
De Huufkes 1, 5511 KC Knegsel, the Netherlands

ABSTRACT

After a short survey of some other measures of perceived image quality, the recently proposed square root integral (SQRI) is described, with special attention to the way in which this measure takes the effect of various display parameters into account.

Experimental data on subjective image quality at varying resolution, addressability, contrast, luminance and display size are compared with predictions by the square root integral. From the comparison it appears that there is a high correlation between perceived image quality and calculated SQRI value.

INTRODUCTION

In the past much effort has been spent to find a measure for the visual resolution quality of a display[1-7]. These measures usually contain the modulation transfer function (MTF) of the display together with the modulation threshold function of the eye. Best known is the so-called modulation transfer function area[2] (MTFA). In this metric the surface area A between the MTF of the display and the modulation threshold curve is used as a measure of visual resolution quality:

$$A = \int_0^{u_{max}} \{M(u) - M_t(u)\} \, du \qquad (1)$$

where u is the angular spatial frequency at the eye of the observer, $M(u)$ is the MTF of the display, $M_t(u)$ is the modulation threshold function of the eye and u_{max} is given by

$$M(u_{max}) = M_t(u_{max}) \qquad (2)$$

However, subtracting $M_t(u)$ from $M(u)$ is not in good agreement with optical convolution rules. Therefore, van Meeteren[3] proposed to replace the difference of $M(u)$ and $M_t(u)$ by the ratio of these quantities:

$$I = \int_0^{\infty} \{M(u)/M_t(u)\} \, du \qquad (3)$$

The inverse of the modulation threshold is usually called contrast sensitivity. Therefore this measure is called integrated contrast sensitivity (ICS). The contrast sensitivity of the eye is the product of the contrast sensitivity of the neuron system and the optical MTF of the eye. So, by using the ratio of $M(u)$ and $M_t(u)$, the display MTF is multiplied by the MTF of the eye, which is in better agreement with optical convolution rules.

Though the ICS is theoretically an improvement over the MTFA, it is still contains a linear function of $M(u)$ in the integrand. This means that the behavior of the eye is implicitly assumed to be linear.

An attempt also to take the non-linear behaviour of the eye into account was made by Carlson and Cohen[6]. In their method, the visibility of displayed information is expressed in a number of just noticeable differences (jnd's) obtained by summation in seven logarithmically spaced spatial frequency channels with a width of factor 2. The discernible levels in the different channels are indicated in so-called discriminable difference diagrams (DDD's). These levels are calculated on basis of a signal detection model, where an increase in signal must overcome a constant fraction of interfering noise. In the model it is further assumed that the number of jnd's increases with the logarithm of $M(u)$ at higher signal levels (Weber's law). However, the agreement between theoretical predictions and measurements is poor. Furthermore, practical application of this method appears to be very complicated. For every change in display parameters, like viewing angle, luminance, scene content and noise level, a different diagram has to be used.

SQUARE ROOT INTEGRAL

These problems are not present in the recently proposed square root integral (SQRI) method[8-10], where the logarithmic summation has been replaced by a logarithmic integration, and the non-linear behavior of the eye is simply taken into account by using the square root of the ratio between display MTF and modulation threshold function. This is justified, because the square root of the modulation depth agrees better with experimental data on noticeable differences at higher signal levels than the logarithmic dependence assumed in the signal detection model of Carlson and Cohen. The square root integral is given by the following equation:

$$J = \frac{1}{\ln 2} \int_0^{u_{max}} \sqrt{\frac{M(u)}{M_t(u)}} \, \frac{du}{u} \qquad (4)$$

where u_{max} is the maximum angular spatial frequency displayed. For a television display, this frequency is determined by the video bandwidth limit of the television signal, but it is also possible to set the upper limit of the integral to infinity and to multiply the display MTF by the MTF of the transmission.

Similarly to the DDD figure of Carlson and Cohen, the display quality given by the square root integral is expressed in just noticeable differences, where 1 jnd is defined as giving a 75% correct response in a two-alternative forced choice experiment. The choice of a well defined jnd unit has the advantage that it eases verification and comparison, and that the user has a good understanding of the practical value of the obtained figure. For a correct interpretation it should be noted that, according to Carlson and Cohen, an increase of 1 jnd must be considered as practically insignificant, an increase of 3 jnd's to be significant, and an increase of 10 jnd's to be substantial. In experiments where the subjective image quality is judged on a quality rating scale, the image quality is often varied over a large

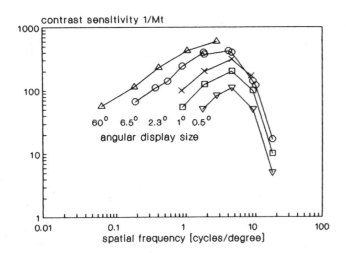

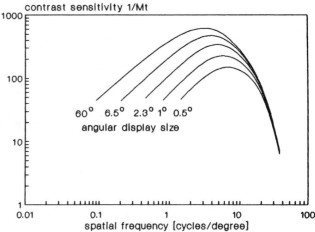

FIG. 1. *Data of the contrast sensitivity 1/M$_t$ as a function of angular spatial frequency at various display sizes, measured by Carlson[13]. Viewing distance 1.9 m. Display luminance 34 mL = 340/π cd/m².*

FIG. 2. *Contrast sensitivity 1/M$_t$ as a function of angular spatial frequency, calculated with Eqs. (5) and (5a) for the conditions of Fig. 1.*

range. In these experiments the number of jnd's per scale unit will generally depend on the size of the scale and the width of the range.

The factor 1/ln2 in front of the integral has been derived under the assumption that the value of J increases by 1 jnd when the square root of M(u)/M$_t$(u) increases by 1 unit in a single spatial frequency channel with a width of a factor 2. This is similar to the assumption made by Carlson and Cohen in their channel model. The factor was also determined by comparing calculation results with measuring data of Carlson and Cohen, which will be treated in one of the next sections. The value found in this way appears to agree within measuring accuracy with the theoretically derived value.

MODULATION THRESHOLD FUNCTION

The square root integral is further defined by using a fixed mathematical expression for the modulation threshold function of the eye, corresponding as closely as possible to that of a standard observer. To obtain such an expression, we analyzed modulation threshold measurements of van Meeteren[3-11]. These measurements were carried out under conditions which most closely approximated normal viewing conditions; i.e., a large viewing angle (17° x 11°), a viewing distance of 4 m, unlimited viewing time, and observation with both eyes and natural pupil. Moreover, the measurements covered a wide range of luminance levels, extending from 10^{-4} to 10 cd/m². We found that the measurements of van Meeteren can very well be described by the following approximation formula:

$$1/M_t(u) = au \exp(-bu) \sqrt{1 + c \exp(bu)} \qquad (5)$$

where
$$a = 440 (1 + 0.7/L)^{-0.2}_{0.15}$$
$$b = 0.3 (1 + 100/L)$$
$$c = 0.06$$
L = display luminance in cd/m²
u = spatial frequency in cycles/degree (cpd)

The parameter "a" determines the low frequency behavior and b and c the high frequency behavior of the contrast sensitivity function. If required, b can be adapted to the visual acuity of a particular observer. Though the maximum display luminance in the measurements was only 10 cd/m², the formula can be used up to much higher values. According to the formula, saturation starts at 100 cd/m². This is in agreement with measurements of van Nes[12] at higher luminance levels.

However, the formula is only valid for pictures viewed at a large viewing angle. At smaller viewing angles, the contrast sensitivity at low spatial frequencies starts to decrease. In order also to incorporate the effect of viewing angle, we analyzed measurements of Carlson[13] on contrast sensitivity at various viewing angles. In these measurements the luminance was constant (34 mL = 340/π cd/m²) and the viewing angle varied from 0.5° to 60°. See Fig. 1. The analysis resulted in the following modification[10] of the low frequency parameter:

$$a = \frac{540(1 + 0.7/L)^{-0.2}}{1 + \dfrac{12}{w(1 + u/3)^2}} \qquad (5a)$$

where w is the angular display size in degrees. For a rectangular picture, an average display size equal to the square root of the picture area should be used. Fig. 2 shows the agreement of calculated modulation threshold values with measurements of Carlson at various viewing angles and Fig. 3 shows the agreement with measurements of van Meeteren at various luminance levels.

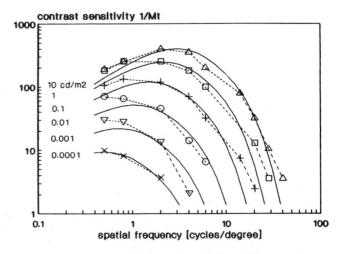

FIG. 3. *Data of the contrast sensitivity 1/Mt as a function of angular spatial frequency at various luminance levels measured by van Meeteren[3,11]. Viewing distance 4m. Angular display size 17° x 11°. The solid lines have been calculated with Eqs. (5) and (5a).*

EFFECT OF VARIOUS DISPLAY PARAMETERS

The square root integral can be used to describe the effect of a large number of parameters on perceived image quality:

1. *Resolution* is taken into account in the display MTF. For the sake of simplicity it is assumed that the resolution is equal in horizontal and vertical direction and constant over the picture area. The spatial frequency of the display MTF is usually expressed in linear units. As the spatial frequency of the MTF in the SQRI formula is expressed in angular units at the eye, this last MTF represents the angular resolution, which still depends on the viewing distance.

2. *Contrast* can also be taken into account in the display MTF. As has been shown in a previous paper[9] contrast loss due to stray light or reflected ambient light causes the modulation depth of all wave components of a luminance pattern to be multiplied by a factor η given by

$$\eta = \frac{L_{av}}{L_{av} + \Delta L} \qquad (6)$$

where L_{av} is the average display luminance and ΔL the increase due to added light. The effect can simply be taken into account by multiplying the display MTF by this factor. This means that J is multiplied by $\sqrt{\eta}$. Contrast changes due to changes of gamma could be treated in a similar way. However, the relation between gamma and η has still to be investigated.

3. *Addressability* is taken into account in the upper limit of the integration. If the addressability is expressed in the number of addressed pixels N over the display width,

$$u_{max} = \tfrac{1}{2} \, N/w \qquad (7)$$

where w is the angular display size.

4. *Luminance* is taken into account in the modulation threshold function. In normal pictures a range of luminances occur. According to Westerink and Roufs[14], the quality impression of a picture is largely determined by the high luminance parts. For pictorial scenes and normal TV pictures we found that the most representative luminance value in this respect is twice the average display luminance. Therefore, the value of L in Eqs. (5) and (5a) should be chosen equal to twice the average luminance in these cases.

5. *Display size* is also taken into account in the modulation threshold function. At a constant viewing distance, the angular display size in the formula depends only on the linear display size.

6. *Viewing distance* has an effect on angular display size, as well as on angular resolution. At a larger viewing distance the angular size of a display decreases, but the angular resolution at the eye increases. This causes opposit effects on the perceived image quality, which makes that there is an optimum viewing distance for a given value of the other display parameters.

7. *Noise* should be taken into account in the modulation threshold function. However, the effect of this parameter has still to be further investigated.

EXPERIMENTAL VERIFICATION

Predictions of the square root integral can be verified in two ways:

1. By experiments where only one display parameter is varied and the required change for a just noticeable difference is measured.

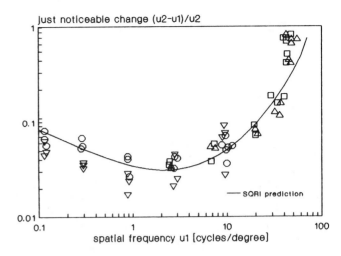

FIG. 4. *Data of the just noticeable fractional change in bandwidth $(u_2-u_1)/u_2$ as function of the original bandwidth u_1, measured by Carlson and Cohen[6]. The bandwidth is defined by the angular spatial spatial frequency where the MTF value of the display has decreased to 0.5. Average display luminance 35 mL = $350/\pi$ cd/m². The solid line has been calculated with the SQRI model.*

2. By experiments where two parameters are varied and the subjective image quality is measured. In this case one parameter can be used to compare the subjective quality scale with the jnd scale and the other to verify the prediction for the combined effect of both parameters.

In this connection it should be remarked that Granger and Cupery[1] showed in their investigation on the optical quality of photographic pictures that there is a linear relationship between subjective quality and just noticeable differences.

RESOLUTION

Carlson and Cohen[6] measured the change in MTF bandwidth needed to obtain a just noticeable change in resolution quality. The measurements were performed at an average display luminance of 35 mL = $350/\pi$ cd/m², with color slide pictures projected on a rear projection screen consisting of two parts. The MTF bandwidth, defined by the 50% value of the MTF, was changed by varying the distance between both parts of the screen. The experiments were carried out at an MTF bandwidth ranging from 0.12 to 50 cpd with two subjects and two different scenes, viewed under an angle of 25.6° (except for the measurements at larger bandwidths, which were performed at a viewing angle of 8.4°).

Measuring data and SQRI prediction are given in Fig. 4. Not only the shape of the predicted curve corresponds to the measurements, but also its vertical position. This confirms the correctness of the factor 1/ln2 in front of the integral expression.

CONTRAST

In a previous investigation the author measured the change in contrast needed to obtain a just noticeable change in image quality[9]. The experiments were carried out using a 625-line color TV monitor with a display size of 52 x 39 cm. The displayed image was a still picture of a street with traffic, buildings, trees and clouds. The picture was viewed at a distance of 2.64 m. Only a single picture and a single observer was used. The video signal was obtained from a flying spot scanner with a slide

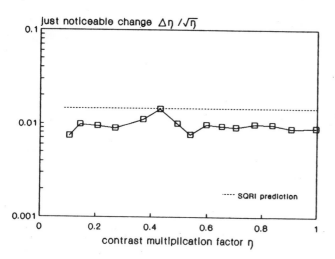

FIG. 5. *Data of $\Delta\eta/\sqrt{\eta}$ as a function of the contrast multiplication factor η, measured by Barten[9]. $\Delta\eta$ corresponds to a just noticeable change. Average display luminance 100 cd/m². According to the SQRI model $\Delta\eta/\sqrt{\eta}$ should be constant. The dotted line gives the SQRI prediction.*

picture of the scene. The contrast was changed electronically by varying the amplitude of the signal at a constant level of the average display luminance (100 cd/m²). The contrast factor η by which the modulation depth of the image was multiplied in this way was measured on the displayed picture. The factor was varied in 15 steps from 0.1 to 1.

Fig. 5 shows the measuring results in a plot of $\Delta\eta/\sqrt{\eta}$ as a function of η, where $\Delta\eta$ corresponds to a change of 1 jnd. From equation (4) it can be derived[9] that $\Delta\eta/\sqrt{\eta}$ should be constant and equal to 2/J (the small effect caused by the luminance change of the higher light levels at constant average luminance can be neglected). The measurements show that this quantity is indeed constant. However, the measured value is smaller than expected: 1 x 10^{-2} instead of 2/J = 1.46 x 10^{-2}. The difference is probably caused by non-linearities of the signals used in the experiment.

RESOLUTION AND ADDRESSABILITY

Knox[15] measured the subjective image quality of machine part line drawings and text images on an 8 point scale. The line drawings were shown to 10 engineers and the text images to 10 secretaries. The pictures were displayed on a monochrome 19 inch CRT monitor at a viewing distance of 22 inches and at a display luminance of 24 cd/m². A factorial combination was made of 4 levels of resolution (50% spot width 5, 10, 15 and 20 mils) and 4 levels of addressability (100, 120, 150 and 200 lines per inch). The addressability was varied by changing the deflection amplitude. However, in this way the size and the luminance of the displayed pictures were varied too.

Fig. 6 shows a linear regression between measured subjective image quality and calculated SQRI value. In the calculation the change in picture size and luminance due to the change in addressability was taken into account. Furthermore, it was also taken into account that the average size of the text images was smaller than that of the line drawings (170 mm instead of 240 mm at the lowest addressability[9]). From the regression analysis it appears that 1 unit of the subjective image quality scale corresponds to 5.1 jnd's. The correlation coefficient amounts to 0.904 and the standard deviation of the data

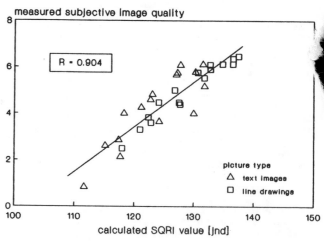

FIG. 6. *Linear regression between measured subjective image quality and calculated SQRI value for measured data from Knox[15] on monitor pictures of line drawings and text images with different resolution and addressability. Correlation coefficient R = 0.904.*

points from the regression line is 0.6 scale point, or 3.2 jnd's. If only the data of the line drawings are used, the correlation is much better. In that case the correlation coefficient amounts to 0.971 and the standard deviation is 0.3 scale point, or 1.5 jnd's.

LUMINANCE AND PICTURE SIZE

Van der Zee and Boesten[16] measured the subjective image quality of projected slides at various luminances and picture sizes on a 5 point scale. The measurements were performed with square color pictures of 5 different complex scenes. They projected simultaneously two images of the same scene, only differing in luminance and size, with the aid of two slide projectors. The viewing distance was constant (2.9 m) and the viewing angle was varied by using copies of the slides in 4 different sizes and by placing each projector at a different distance. In this way the picture size varied in 7 steps from 0.21 m to 1.02 m, corresponding to a viewing angle ranging from 4.2° to 20°. The luminance of each projector was varied by using neutral density filters. In this way the average luminance varied in 6 steps from 7 cd/m² to 50 cd/m². The pictures were shown in random order to 29 subjects.

Fig. 7 shows a linear regression between measured subjective image quality and calculated SQRI value. From the regression analysis it appears that 1 unit of the subjective image quality scale corresponds to 14.4 jnd's. The correlation coefficient amounts to 0.993 and the standard deviation of the data points from the regression line is 0.09 scale point, or 1.4 jnd's.

RESOLUTION AND PICTURE SIZE

Westerink and Roufs[14] measured the subjective image quality of projected slides at various resolutions and viewing angles on a 10 point scale. The measurements were performed at an average display luminance of 30 cd/m² with square color pictures of 5 different complex scenes of which 2 were the same as used in the above mentioned investigation of van der Zee and Boesten. In this case only one projector was used. The viewing distance was constant (2.9 m) and the viewing angle was varied by using copies of the slides in 4 different sizes. In this way the picture size varied in 4 steps from 0.24 m to 0.92 m, corresponding to a viewing angle ranging from 4.7° to 18°. The resolution was varied in 7 steps

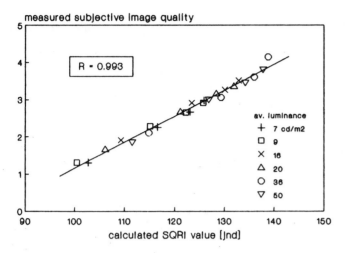

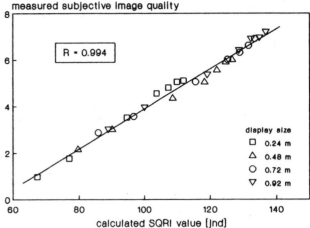

FIG. 7. Linear regression between measured subjective image quality and calculated SQRI value for measured data from van der Zee and Boesten[16] on projected color slide pictures with different luminance and size. Viewing distance 2.9 m. Correlation coefficient R = 0.993.

FIG. 8. Linear regression between measured subjective image quality and calculated SQRI value for measured data from Westerink and Roufs[14] on projected color slide pictures with different resolution and size. Viewing distance 2.9 m. Average display luminance 30 cd/m². Correlation coefficient R = 0.994.

from 2.7 cpd to 38 cpd (50% value of Gaussian MTF) by defocusing the projector lens with the aid of a stepper motor. The pictures were shown in random order to 20 subjects, all with a visual acuity of at least 1.0.

Fig. 8 shows a linear regression between subjective image quality and calculated SQRI value. From the regression analysis it appears that 1 unit on the subjective image quality scale corresponds to 11.6 jnd's on the SQRI scale. The correlation coefficient amounts to 0.994 and the standard deviation of the data points from the regression line is 0.19 scale point, or 2.2 jnd's.

In Fig. 9 the SQRI values of this experiment are plotted as a function of the number of pixels per display width, resolved with an MTF value larger than 0.5. This number is given by

$$N = 2 \ w \ u_{0.5} \qquad (8)$$

if w is the angular display size and $u_{0.5}$ the spatial frequency where the MTF value has decreased to 0.5. The figure shows, that for a given number of pixels, a larger picture size not allways means a better quality. The curves indicate, furthermore, that there is an optimum display size for each value of N at the given viewing distance. This also means that for a given picture size there will be an optimum viewing distance for each value of N. This subject will be treated in more detail in the following section.

VIEWING DISTANCE

The optimum viewing distance was investigated already more than 30 years ago by Jesty[17]. He measured the optimum viewing distance for projected color pictures of different size and resolution. The aspect ratio of the pictures was 4 : 3. In the experiments 4 different color slides and 9 observers were used. In this way each measuring point was the average of 36 observations.

In a first experiment the optimum viewing distance was determined at 4 different display heights, ranging from 9 inch to 26 inch and two levels of total resolution ("sharp" and "defocused"). The luminance was kept constant at a level of 20 ftL = 69 cd/m² for the highlights by using neutral density filters. From this experiment it appeared that the ratio between optimum viewing

distance and picture size was constant at constant total resolution. Such a behavior is in agreement with the SQRI model, as the square root integral contains only angular dimensions.

In a second experiment additional levels of total resolution were created by projecting different parts of the original pictures. In this procedure it is assumed that the sharpness does not vary over the picture and that the total resolution is proportional to the projected area of the original picture. Furthermore, there were also measurements carried out at 1/4 and 4 times the original luminance.

Fig. 10 shows a linear regression between the measured optimum viewing distance and the value calculated with the square root integral for the "sharp" pictures. In this figure the viewing distance is expressed in units of picture height. For the calculations one assumption had

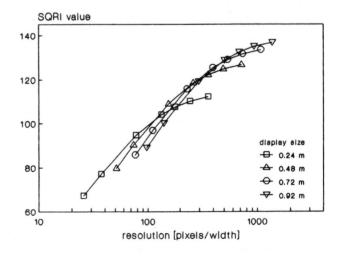

FIG. 9. SQRI values calculated for the measurements of Westerink and Roufs given in Fig. 8, as a function of the number of pixels per display width resolved with an MTF value larger than 0.5. Viewing distance 2.9 m. Average display luminance 30 cd/m².

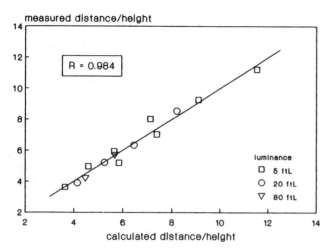

FIG. 10. *Linear regression between measured and calculated optimum viewing distance for measured data of Jesty[17] on projected 4 : 3 color slide pictures with different resolution and luminance. The optimum viewing distance is expressed in units of picture height. Correlation coefficient R = 0.984.*

to be made with respect to the resolution of the pictures, as exact data were lacking. It was assumed that at complete projection the picture had a resolution of 540 pixels per display width (defined by the 50% value of a Gaussian MTF). Though the calculations could be fitted to the measurements in this way in a single point, the behavior of the other points shows that measurements and calculations further completely agree. The correlation coefficient amounts to 0.984 and the standard deviation of the viewing distance from the regression line 0.39 times the picture height. This deviation corresponds to a difference of less than 0.1 jnd with respect to the maximum SQRI value. This amazingly small value means that the resolution quality of a picture can very accurately be determined by measuring the optimum viewing distance.

CONCLUSIONS

From the agreement between experimental data and theoretical predictions at variation of resolution, addressability, contrast, luminance, picture size and viewing distance, it appears that the square root integral is a good measure for perceived image quality. By the use of a mathematical equation for the contrast sensitivity of the eye the application is relatively simple.

REFERENCES

1. E.M. Granger and K.N. Cupery, An Optical Merit Function (SQF), Which Correlates with Subjective Image Judgments, Photographic Science and Engineering, Vol. 16, No.3, pp. 221-230, May-June 1972.

2. H.L. Snyder, Modulation Transfer Function Area as a Measure of Image Quality, Visual Search Symposium Committee on Vision, National Academy of Sciences, Washington D.C., 1973.

3. A. van Meeteren, Visual Aspects of Image Intensification, Thesis, University of Utrecht, The Netherlands, 1973.

4. G.C. Higgins, Image Quality Criteria, Journal of Applied Photographic Engineering, Vol. 3, No. 2, pp. 53-60, 1977.

5. H.L. Task, A.R. Pinkus and J.P. Hornseth, A Comparison of Several Television Display Image Quality Measures, Proc. of the SID, Vol. 19, No. 3, pp. 113-119, 1978.

6. C.R. Carlson and R.W. Cohen, A Simple Psycho-physical Model for Predicting the Visibility of Displayed. Information, Proc. of the SID, Vol. 21, No. 3, 1980.

7. R.J. Beaton, Quantitative Models of Image Quality, Proc. Human Factors Soc. 27th Annual Meeting, pp. 41-45, 1983.

8. P.G.J. Barten, The SQRI Method: a New Method for the Evaluation of Visible Resolution on a Display, Proc. of the SID, Vol. 28, No. 3, pp. 253-262, 1987.

9. P.G.J. Barten, Evaluation of CRT Displays with the SQRI Method, Proc. of the SID, Vol. 30, No. 1, pp. 9-14, 1989.

10. P.G.J. Barten, Effect of Picture Size and Definition on Perceived Image Quality, Record 1988 Int. Display Research Conf., pp. 142-145, Oct. 1988.

11. A. van Meeteren and J.J. Vos, Resolution and Contrast Sensitivity at Low Luminances, Vision Res., Vol. 12, pp. 825-833, 1972.

12. F.L. van Nes, Experimental Studies in Spatiotemporal Contrast Transfer by the Human Eye, Thesis, University of Utrecht, The Netherlands, March 1968.

13. C.R.Carlson, Sine-Wave Threshold Contrast-Sensitivity Function: Dependence on Display Size, RCA Review, Vol. 43, pp. 675-683, Dec. 1982.

14. J.H.D.M. Westerink and J.A.J.Roufs, A Local Basis for Perceptually Relevant Resolution Measures, SID Digest, Vol. 19, pp. 360-363, May 1988.

15. S.T. Knox, Resolution and Addressability Requirements for Digital CRT's, SID Digest, Vol. 18, pp. 26-29, May 1987.

16. E. van der Zee and M.H.W.A. Boesten, The Influence of Luminance and Size on the Image Quality of Complex Scenes, IPO Annual Progress Report, Vol. 15, pp. 69-75, 1980.

17. L.C. Jesty, The Relation between Picture Size, Viewing Distance and Picture Quality, Proc. IEE, Vol. 105B, pp. 425-439, Feb. 1958.

Visual multipoles for quantifying raggedness and sharpness of images

Stanley Klein and Thom Carney

School of Optometry, UC Berkeley

At the recent SPIE meeting on Human Vision, Visual Processing and Digital Display (1989) it was clear that many display engineers are interested in the properties of the human visual system. Knowledge about human performance is being used both for improving a metric to measure image quality and also to improve data compression algorithms. It was apparent from that meeting that the languages used by the display engineers and the vision researchers have not fully meshed. The present meeting on applied vision provides an excellent forum for further discussion and cross-fertilization between the groups and the present paper is offered in that spirit. This paper is an outgrowth of our psychophysical experiments on the detection of blur and the detection of misalignment as a function of contrast. In the Discussion we will describe the relevance of this research for perceived image quality and for compression algorithms.

The work of Hamerly and colleagues at Xerox provide a good background for our studies. In a pair of articles published in 1981 they examined how the characteristics of the human visual system can provide relevant information for establishing a metric of image quality. Hamerly and Dvorak (1981) in a paper titled "Detection and discrimination of blur in edges and lines" found that the blur (resolution) threshold of a high contrast sharp edge was about 25 sec. They were surprised to find that when the initial edge is blurred, the blur threshold can be reduced to as low as 5 - 10 sec of arc. They also examine the blur threshold for different contrasts and for different blur profiles. We believe that our multipole formalism is able to explain all of their data in a concise, model free manner. Another reason for referring to the Hamerly et al paper is that its 13 references all refer to the display engineering and image quality literature thus providing psychophysicists with an entry into that field. The second paper, by Hamerly and Springer (1981), is titled "Raggedness of edges" and examines the relevance of channel models for making judgements about the straightness of edges. These two papers by Hamerly and colleagues seem to be about two different aspects of image quality since they involve two different psychophysical tasks. The detection of blur is a resolution task whereas the detection of raggedness is similar to a vernier task. One of the outstanding puzzles of spatial vision is how to relate vernier thresholds that are in the hyperacuity range of 3 - 5 sec to resolution thresholds that are about 10 times higher.

The multipole formalism allows blur detection and raggedness detection to be handled within a single formalism. Three visual multipoles (edge, line, and dipole) are relevant to the present paper. The line stimulus is the derivative of the edge and the dipole is the derivative of the line. The top half of Figure 1 shows how vernier acuity of a line can be understood as the addition of a dipole to a line. The bottom half of the figure shows how blur detection of an edge (edge resolution) can be understood as the addition of a dipole to an edge. Instead of measuring vernier acuity in terms of seconds of arc, it can be measured in terms of the threshold dipole strength that is needed to detect the vernier offset. Similarly, edge blur can be expressed in units of the dipole strength that is needed to produce a just visible blur.

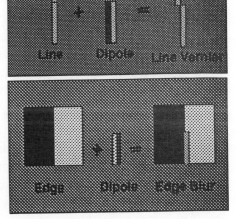

Figure 1

Mathematical details. The quantitative connection between vernier offset, resolution size and the multipole strength requires a detailed development of the multipole formalism (Klein 1990). Here we merely report the results without justification. For brevity we focus on the dipole test pattern. The dipole moment has units of $\%\min^2$, meaning the contrast times the line width (for one of the lines comprising the dipole) times the separation between the two opposite polarity lines. Instead of expressing resolution and vernier acuity as purely spatial measures (sec of arc) our formalism in terms of multipole moments assumes that the contrast of the stimulus is also important for setting the threshold.

The connection between line-vernier offset, δ, line strength, l, and dipole moment, d, is:
$$d = l\,\delta, \tag{1}$$
where l is the product of the contrast times the width of the line. The connection between the resolution threshold, r, edge strength, c, and dipole moment, d is:
$$d = 1/8\ c\ r^2 \tag{2}$$
where c, the contrast of the edge is <u>twice</u> the Michaelson contrast. This contrast definition is forced by the multipole formalism. The blur extent, r, is the standard deviation of the filter (such as a Gaussian) that is convolved with the edge to produce the blur.

Figure 2

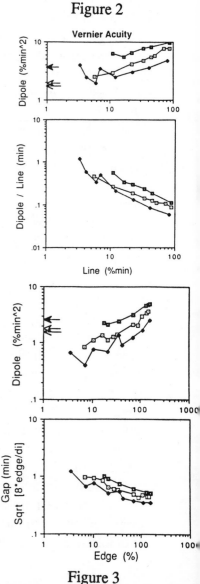

<u>Experimental methods and results.</u> A VENUS pattern generator was used to present stimuli on a Tektronix 608 monitor. The mean luminance was about 100 cd/m^2, the pixel size was .3 min, and the field size was 1.3 deg. A method of constant stimuli with typically 5 stimuli was used. The observers gave a rating scale response in which there were as many rating categories as there were stimuli. Threshold was defined at the point where d' = 1.

Vernier thresholds for three observers are shown in Fig. 2. The abscissa is the line moment (contrast x width) of the pedestal line. The lower panel has thresholds expressed in sec of arc. In the upper panel the same thresholds are shown but they are now converted to the threshold dipole moment which is the offset threshold in min times the pedestal line moment (the abscissa) in %min. Thus the units of the dipole moment are $\%\min^2$. The arrows at the left are the dipole detection thresholds when there is no pedestal. The ability to plot this new data is the justification for the multipole approach. The detection threshold is a measure of the signal to noise ratio for the vernier task. Rather than normalizing mechanism sensitivity using the CSF as is done by most vision researchers, we have chosen to normalize the sensitivity using a stimulus that is similar to that which is found in the vernier task. It is gratifying that the dipole moment for the vernier task is approximately the same as the dipole moment in a detection task.

Fig. 3, shows similar data for an edge blur discrimination task. The abscissa is the edge contrast. In the lower plot the resolution threshold is expressed in sec of arc, where the angular size corresponds to the standard deviation of the edge blur. In the upper plot the thresholds are transformed to dipole moment units. The arrows again show the dipole thresholds. The blur thresholds are <u>better</u> than the

Figure 3

dipole detection threshold. This is different than what was found for vernier acuity and is at first surprising for two reasons. 1. It implies that resolution is better than vernier acuity. 2. It implies the visibility of a dipole is enhanced by the presence of a pedestal.

This latter point is similar to the facilitation found in contrast discrimination (Stromeyer and Klein, 1974) that has been come to be known as the dipper function (Legge and Foley 1980). It gets this name because for pedestals slightly above threshold the contrast discrimination is about half of the threshold value. Fig. 4, shows the dipper function for contrast discrimination together with the vernier and resolution data from the upper panels of Figs. 2 and 3. This comparison shows that the edge pedestal in the resolution task produces a facilitation quite similar to that shown by the dipole pedestal in the contrast discrimination task. The vernier task does not show the facilitation.

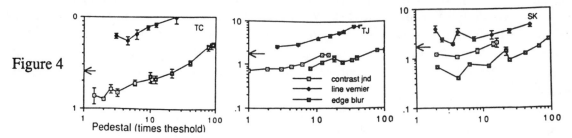

Figure 4

Discussion. In our approach, hyperacuity and resolution thresholds are not to be measured by reporting thresholds in purely spatial units. Rather the task is to be understood in terms of a test-pedestal paradigm, in which the test pattern is that pattern which must be added to the pedestal to produce the blur (for a resolution task) or the offset (for a hyperacuity task). Thresholds are reported as the strength of the test pattern. For localized pedestals such as edges and lines the test patterns are higher order multipoles and their strength is quantified by the multipole moments discussed earlier.

Our results offer a new perspective on the relationship between hyperacuity and resolution. In all earlier approaches the challenge was to explain why hyperacuity was so much better than resolution acuity. Now that the thresholds are expressed as multipole moments, so that the effect of contrast is taken into account, we find that resolution thresholds are slightly better than hyperacuity thresholds. It is easy to offer suggestions for why this might be. The vernier task undoubtedly uses mechanisms that are oriented with respect to the dipole. These mechanisms would therefore not have good overlap with the dipole so the thresholds would be expected to be worse than for contrast discrimination. The resolution task on the other hand would use mechanisms aligned with the dipole that could be quite sensitive to detecting the dipole test pattern. For large pedestal strengths we would also expect that the resolution task should have lower thresholds than the contrast task because for the contrast task the test pattern is identical to the pedestal, so all the detection mechanisms would be masked. For the resolution task the test and pedestal are different so there will be high spatial frequency mechanisms sensitive to the dipole that are not greatly masked by the edge.

Our data has some relevance to questions of how to quantify image quality and of how the image information might be compressed. The bottom portions of Figs. 2 and 3 show that as contrast is increased the visibility of the blur and of the offset is decreased. Thus a low contrast blurred and jagged image could still have good image quality and could withstand greater compression (sparser sampling) than a high contrast image.

As discussed in the introduction, Hamerly and Dvorak (1981) found resolution thresholds of 25 sec for a sharp edge but thresholds as low as 5-10 sec for blurred edges. Thus the discrimination of edge blur can be facilitated by the presence of blur. This finding has striking consequences for the image quality issue since it seems to imply that in order to maintain high resolution one should sample 5 times what one had previously thought (5 sec rather than 25 sec). We would like to point out that in terms of our approach the facilitation was expected. A pedestal with blur corresponds to a task of detecting a change in the dipole moment in the presence of a pedestal dipole moment. Eq. 2 can be used to calculate the Δd needed to detect the change in dipole moment:

$$\Delta d = 1/4 \, c \, r \, \Delta r. \tag{3}$$

If the Δd is constant then Δr (the blur Hamerly measured) is inversely related to r, the pedestal blur, in agreement with Hamerly's findings. Another way to say this is that if image quality is measured in terms of dipole threshold then one wouldn't have the disturbing finding that the image quality constraint is stricter for blurred images. One must simply ensure that the display has adequate spatial resolution (about .5 min) and adequate contrast steps (about 1% increments) to be able to display all possible dipoles.

This report only skimmed the surface of how multipoles can be used in studying vision. The SPIE conference publication (Klein , 1989) presented an overview of the multipole formalism, how it is related to the sinusoid formalism (CSF), how multipoles can be generated on video terminals, how the visibility of pairs of multipoles depends on their separation, and how multipoles provide a multiplicity of advantages for the clinical assessment of visual performance. It is our belief that the multipole formalism may also prove useful for display engineers.

References.

J.R. Hamerly and C. A. Dvorak, "Detection and discrimination of blur in edges and lines," J. Opt. Soc. Am. **71**, 448-542 (1981).

J.R. Hamerly and R. M. Springer, "Raggedness of edges," J. Opt. Soc. Am. **71**, 285-288 (1981).

G. E. Legge and J. M. Foley, "Contrast masking in human vision," J. Opt. Soc. Am. **70**, 1458-1471 (1980).

S. A. Klein, "Visual multipoles and the assessment of visual sensitivity to displayed images," SPIE conference held in Los Angeles (1989).

S. A. Klein, "Multipole sensitivity of the visual system," manuscript in preparation (1990)

C. F. Stromeyer III and S. A. Klein, "Spatial frequency channels in human vision as asymmetric (edge) mechanisms," Vision Res. **14**, 1409-1420 (1974).

Distortion Metrics for Image Coding Using Monochrome and Color Human Visual Models

Scott E. Budge

Utah State University, Department of Electrical Engineering
Logan, Utah 84322

Abstract

The use of a human visual model based distortion measure for image coding and evaluation is described. Test results show that when an image is encoded using the visual model, the measured quality correlates better with visual quality when the measurements are made within the model, and better images are produced.

Paper Summary

Since communication channels are limited, it has become necessary to reduce the bandwidth required to transmit and receive digital images for many applications. Several different methods have been employed to compress image data, and often these methods require a measure of the distortion caused by the encoding process to evaluate the success of the method. A good example of this is the technique known as vector quantization. This coding method requires a measure of distortion to implement the encoder, and the quality of the output is often evaluated by the same measurement.

This paper describes the use of a monochrome and color visual model to modify the traditional mean-square-error distortion metric used in vector quantization. The image is transformed into a "perceptual space," and the distortion computations made by the vector quantizer encoder are performed on the new image. The quality of the image can then be measured in the perceptual space to determine how well the algorithm has encoded the image.

The vector quantization (VQ) algorithm evaluated in this paper is a slight variation of the standard VQ, designed to be more robust to variations in overall image intensity. Known as the Mean/residual Vector Quantizer (MRVQ), this algorithm separates the mean intensity and the residual textural information of each vector and encodes them separately using optimal codebooks [1]. The resulting product code is then transmitted.

It has long been known that the intensity values that are digitized from an image produce a histogram which is highly skewed to the lower intensities. From a coding standpoint, it is much more desirable to have a uniform distribution of pixel values so that the

VQ algorithm is more effective. One of the simpler monochrome visual models, proposed by Stockham, has as its first step a logarithmic mapping which is an attempt to model the intensity-to-brightness conversion of the visual system [2]. I has the effect of shifting the histogram of the pixel data so that it is more uniform.

The second step in the model is a linear space-invariant filter which models the lateral-inhibition process in the retina. This enhances the high frequency spatial components of the image and provides the VQ encoder an image rich with edges. Since vector quantization has problems with edges, this step emphasizes them so that they may be better encoded.

The color model is an extension of the monochrome model by Stockham, and uses the same basic modeling steps [3]. This model also provides for the chromatic response of the retina, and converts the three color channels into an intensity and two chrominance channels. The chrominance channels are then subsampled to provide greater compression, and the three channels are quantized using the MRVQ.

Once the image has been transformed, the distortion between vectors extracted from the image and the VQ codebook is measured using a mean-square-error (MSE) criterion. Since the image is in a new data space, the measure is no longer the same as the MSE in the intensity space. This new measurement is used to determine the best codevector to represent the source vector. The index of this codevector in the codebook is then sent, along with the encoded mean, to the decoder for reconstruction of the image.

The final step in the process is to perform the inverse of the visual model at the decoder so that the image can be properly viewed. This image is the image that is displayed on the receiver output device and evaluated visually.

Tests were run to determine how well the new distortion measurement performs. Several images were encoded using both the intensity distortion metric and the perceptual space metric, and the results were compared by visual evaluation and distortion measurements in both the intensity and perceptual spaces. These evaluations have shown that for both the monochrome and color models, the visual quality of the encoded images were better when the perceptual space metric was used, and the perceptual space MSE correlated more closely to the perceived quality of the images. This suggests that a perceptual space metric must be used to have a useful measurement of coded image quality.

It should be noted, however, that the visual model used in this work is a simplified model. The correlation between visual image quality and distortion measurement were better, but some images were encoded which did not follow the trend. Thus, a better model could produce a higher correlation. The model presented here was selected because of its relative ease of implementation in a signal processing environment.

References

[1] Richard L. Baker, *Vector Quantization of Digital Images*, Ph.D. dissertation, Stanford University, June 1984.

[2] Thomas G. Stockham, Jr., "Image processing in the context of a visual model", *Proceedings of the IEEE*, 60(7) pp.828–842, July 1972.

[3] O. D. Faugeras, "Digital color image processing within the framework of a human visual model", *IEEE Transactions on Acoustics, Speech, and Signal Processing*, 27 pp.380–393, Aug 1979.

Psychophysical rating of image compression techniques[*]

Charles S. Stein,† Andrew B. Watson,‡ and Lewis E. Hitchner†

† University of California, Santa Cruz Computer and Information Sciences
Applied Sciences Bldg., UCSC, Santa Cruz, CA 95064

‡ Vision Group, NASA Ames Research Center
mail stop 239-3, Moffet Field, CA 94035

1. INTRODUCTION

To rate lossy image compression techniques in terms of transmission rate, distortion induced by the compression and reconstruction process, must be measured in a way that correlates well with human perception of the distortion. Many researchers agree that the common computational distortion metrics mean-square error (MSE) and root mean-square error (RMSE) do not correlate well with human subjective judgment.[1-8] In an effort to find a fidelity metric that corresponds more closely to human fidelity perception, some researchers have experimented with other measures such as the MSE weighted by properties of the human visual system (HVS).[2,5,6,9-14] No standard computational fidelity metric has been adopted by all researchers, and because they are computationally simple, MSE based measures remain the *de facto* metric. Most image fidelity testing with humans has been performed using different methods or under conditions that vary from lab to lab so that it is difficult to compare one researcher's results with those of another. Statements in the literature such as ". . . an average of 0.26 bits/pixel can be achieved with very little perceivable degradation" or "Good Quality images are achieved with as low as .5 bit/pel"[15] make it nearly impossible to compare image compression methods presented in different papers.

Three types of fidelity measures exist. *Computational* includes all MSE-based measures, *subjective perceptual* is based on rating scales and ordering images based on some criterion, and *objective perceptual* which involves the use of psychophysical testing methods.[16] The purpose of most previous image fidelity testing using subjects, has been to test computational image fidelity metrics. Commonly, researchers have compared various computational metrics and identifed the best as the one with the highest coefficient of correlation with results from a subjective perceptual evaluation. The most popular subjective perceptual evaluation methods are the CCIR quality and impairment scales in which descriptive phrases are associated with numerical scores. The problem with these methods is that the preworded rating scale is not criterion free, and they do not produce a unique repeatable measure of fidelity that can be reproduced in other labs.

Rating different image compression techniques can be accomplished by using the same *perceptual* test and *constant* viewing conditions. Watson devised a psychophysical test that identifies that identifies a just-noticeable-distortion level from a set of images varying in distortion.[17] This test is an objective, criterion free, perceptual evaluation method. We used this test to identify which images, produced by varying the degree of distortion in different image compression methods, were perceived to have equally visible distortion at a level where distortion is just perceptible. We tested five different digital image compression methods on a single, monochrome original image. The five methods tested used the lossy compression algorithms: Adaptive Block Truncation (BTC),[18-22] Block Discrete Cosine Transform (DCT),[23,24] with and without an HVS scaling factor,[25] Laplacian Pyramid,[26] and Cortex Transform[17,27] coding. The image compression methods were rated by comparing the bit/pixel rate of images identified as having the same subjective distortion level.

2. METHODS

2.1. Apparatus

An Adage RDS-3000 raster graphics system computer was used to control the entire experiment. It drove a Grinnell Systems monochrome monitor used to display images, and coordinated synthetic voice response with observer input on the terminal keyboard. The display monitor was calibrated such that pixel values were linearly proportional to the luminance. Images were displayed at a contrast of 100%, with a mean luminance of 100 cd/m^2. The images were displayed at a size of 256 x 256 pixels at a (horizontal and vertical) resolution of 20 pixels/cm and were viewed by subjects with their heads resting on a chin bar at a distance of 91.5 cm (7.15 times image height). These conditions created a viewing angle of 8° and an image Nyquist frequency of 16 cycles/degree which allowed all spatial frequencies in the image to be seen. A more detailed description on the method of image display and calibration is described by Watson et al.[28]

---

[*] Summary from a paper by the same title published in SPIE Proceedings of the Human Vision, Visual Processing, & Digital Display conf., Vol. 1077, 1989.

2.2. Subjects

Subjects were paid untrained observers, men and women between the ages of 17 and 40. All had normal or corrected-to-normal vision. Because subjects develop strategies quickly, such as searching for particular artifacts of the compression method, their performance usually improves after the first testing session. Thus, subjects were considered as belonging to two groups depending on experience. Group 1 consisted of subjects performing the experiment for the first time (naive), while group 2 consisted of subjects who had already performed the experiment once (experienced).

2.3. Experimental design

For each compression method tested, many images were produced by varying the parameters, treatment factors, that control distortion level (i.e. for BTC, blocksize and quantization strength). A complete factorial design was used where all possible combinations of the treatment factors were used. If a method had 2 treatment factors, the images were divided into groups under the first factor that varied the strength of the second factor. Each compression technique has different control parameters that vary the amount of compression and allowable distortion; how these were varied is discussed in the stimuli section.

One testing session consisted of a two-alternative forced-choice staircase experiment performed on all possible combinations of the treatment factors for a particular compression method. Prior to the experiment, the subject had a training session that included viewing one at a time all the images used in the test and a practice session of the experiment of up to 2 minutes long or until he or she felt comfortable with the procedure. The experiment followed immediately. In each trial, the subject signaled with keyboard input when ready, a beep was sounded, and two images were displayed side by side on a mean luminance surround for 2 sec. One image was the original and one the reconstructed image with the reconstructed image randomly placed on the left or the right. The subject's task was to choose which was the coded image. A response of right or left was entered on the keyboard using the 1 or 2 key on the numeric keypad and a voice synthesizer responded with "correct" or "wrong" appropriately. Each group of images under the first treatment factor consisted of 64 trials. Each compression method was tested by naive and experienced subjects.

The *QUEST* staircase method[29] is used to select which image, or level of the second factor, under a particular condition of the first factor is to be tested next. Correct responses cause the next trial for that condition to be weighted towards an image with less distortion while incorrect responses cause a weighting towards one with more distortion. Trials of different conditions of the first factor are randomly interleaved so that the subject is unaware of which condition is being tested. The trials are designed to converge on a distortion level that yields 82% correct responses.

2.4. Stimuli

This experiment was performed using the "Lenna" image[30] which contains a wide range of spatial frequencies. We used a reduced, 256 x 256, version of the original 512 x 512 image. The image was reduced by removing the high spatial frequency octave from the original in the Fourier frequency domain, subsampling, then inverse transforming the result. The experiment was performed on one image because of the time involved in fully testing one image and the space management problem with approximately 10 M bytes needed for all the test images.

In all compression methods except Cortex transform coding, quantization strength, Q, was used as the second control parameter, or blocking factor, that varied distortion as well as bit-rate. Cortex transform coding used only one factor, quantization strength. A logarithmic relationship between quantization strength and the number of quantization bins was derived for each method, and this parameter, Q, was incremented in either octave or half octave steps. The value of Q has no necessary relation between compression methods. A range of values for Q was chosen such that when it was varied, each group of reconstructed images produced a range of 50% - 100% correct responses by subjects. This guaranteed that the 82% correct responses point could be interpolated with reasonable accuracy. The range of values for Q was found through initial testing by one of the authors.

3. RESULTS

Results of naive and experienced subjects were analyzed separately. A Weibull function was fitted to the data for each test session and the point where 82% correct responses occurred was interpolated along this curve.[31] For each group and each condition of the first blocking factor, the quantization strength associated with 82% correct responses was averaged and considered as 1 just-noticeable-difference (JND) from the original image.† Within each group, all images identified at the 1 JND level are considered as being perceptually equivalent for comparison purposes. Among the different conditions of each compression method, results were close. To show the general relationship between methods, results from all conditions were

---

† Because the interpolation is not accurate outside the range of quantization strengths used, results indicating a strength above or below those tested were set to the minimum or maximum test strength appropriately. This effectively bounds the average score within the testing range. This occurred very rarely.

averaged and plotted in the graphs of Figure 1 for entropy vs. method and SNR vs. method.

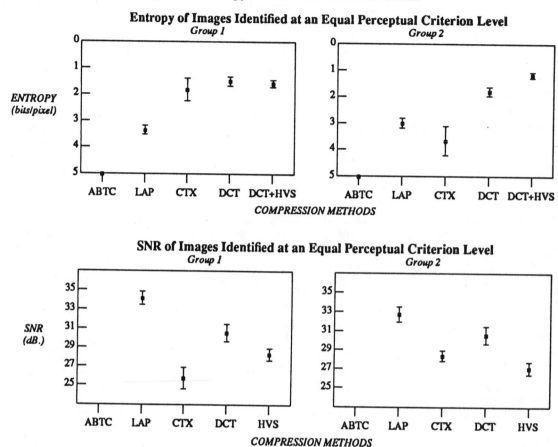

Figure 1: Plots of entropy and SNR vs. compression method for mean over all conditions tested. Error bars indicate +/- 2 standard errors. ABTC did not produce a perceptually equivalent image. SNR is normalized mean-square-error based. LAP is Laplacian pyramid coder, and CTX is Cortex transform coder.

4. DISCUSSION

4.1. Bit-rate of perceptually equivalent images

DCT with and without the HVS scaling factor consistently had the lowest bit-rate, while Laplacian pyramid coding was consistently high, and Cortex transform coding had an inconsistent bit-rate between groups. ABTC images were almost always discriminated, and did not yield any images at the 1 JND level. Subjects found it very easy to detect the block structure associated with ABTC, even with block sizes of 2 x 2. In group 2, DCT with HVS scaling has a lower bit-rate than DCT alone, but the size of the group for DCT with HVS was small. Thus, it is not clear that the HVS scaling factor used improves performance (as measured by entropy) of the DCT Coder. The inconsistency in the bit-rate results between groups for the Cortex transform coder requires further study. One group 2 subject who scored 62 out of 64 correct responses on the Cortex transform test mentioned using "regular patterns" he saw along the boundaries of the test images to discriminate from the original.

The results of group 2 should be expected to show a better discrimination resulting in images with a higher bit/pixel-rate. This was not true with the DCT using HVS scaling. This could possibly be due again to the small size of group 2 (2 subjects). All compression methods except Cortex transform coding showed very close results between the naive and experienced groups. Because the Cortex transform coding had only 1 set of images tested (64 trials), a naive subject had little time to improve performance, while a naive subject performing the DCT test with 5 sets of images (5*64 trials) had much more time to improve performance.

4.2. HVS based techniques mask distortion better

Of all images identified at an equal perceptual criterion level, the Cortex transform and DCT with HVS scaling images had the lowest SNR. Both these methods give more bits to the spatial frequencies humans are most sensitive to, and less to those we are less sensitive to. These two methods, each based on the HVS, mask distortion (as measured by SNR) better than the other techniques. The Laplacian method had the highest SNR, and although it does vary bit-rate between spatial frequency bands, it does not make use of the nonlinear distribution of values on a particular pyramid level. Performance could likely be improved by the use of a nonlinear or visual quantizer.

4.3. Accumulation of distortion

Results of Laplacian pyramid and Cortex transform coding, both multiresolution techniques, showed bit-rates higher than those published and found using the same testing criterion on individual resolution levels.[17, 26, 32] These methods varied amount of quantization on one resolution at a time to determine the correct strength. Our results, of significantly higher bit-rates than those previously published, imply that the distortion induced by quantization in different resolution levels (separate spatial frequency bands) accumulates. This was also noted by Carlson and Cohen,[33] but has not been examined carefully in the context of image compression.

4.4. SNR as a fidelity metric

As clearly visible in Figure 1, the correlation between normalized MSE-based SNR and the perceptually equivalent images is very poor. MSE is not sensitive to how humans respond to different forms of distortion. A robust fidelity metric would measure these images the same. The perceptually equivalent images identified using our objective testing method provide an excellent opportunity to verify various computational fidelity metrics.

4.5. Results in perspective

Our results are preliminary and are not meant to be interpreted in viewing conditions other than those described here. The monochrome monitor used has a very excellent frequency response. Artifacts in our test image are more visible when the greyscale is gamma corrected for the monitor (as in our test conditions) than with a linear greyscale. Distortion artifacts noticeable on this monitor in our test conditions often cannot be seen on color or other monitors. The testing procedure is strict, and invites the subject to focus in on areas of the image where they expect to see artifacts. Thus, our results reflect a higher bitrate than might be expected by most viewers on average equipment in an average environment.

One important limitation of our results is the use of only one test image. Because of time and storage constraints we were limited in this respect. It is important to find ways to facilitate subjective testing of many images in an objective environment such as the one described here.

The psychophysical testing method presented here is an objective, criterion free, perceptual evaluation method. It allows us to compare, at a perceptually equal distortion level, image compression techniques that induce different forms of distortion. More objective perceptual image fidelity evaluation will lead to a better understanding of both perceivable and unperceivable distortion, and thus to better image fidelity metrics.

5. REFERENCES

1. John R. Parsons and Andrew G. Tescher, "An Investigation of MSE Contributions in Transform Image Coding Schemes," *SPIE*, vol. 66, pp. 196-206, 1975.

2. John O. Limb, "Distortion Criteria of the Human Observer," *IEEE Trans. on Sys., Man, & Cybernetics*, vol. SMC-9, no. 12, pp. 778-793, Dec. 1979.

3. Norman C. Griswold, "Perceptual coding in the cosine transform domain," *Optical Engineering*, vol. 19, no. 3, pp. 306-311, May/June 1980.

4. Anil K. Jain, "Image Data Compression: A Review," *Proc. of the IEEE*, vol. 69, no. 3, pp. 349-389, March 1981.

5. Norman B. Nill, "A Visual Model Weighted Cosine Transform for Image Compression and Quality Assessment," *IEEE Trans. on Comm.*, vol. COM-33, no. 6, pp. 551-557, June 1985.

6. Frank X. J. Lukas and Zigmantas L. Budrikis, "Picture Quality Prediction Based on a Visual Model," *IEEE Trans. on Comm.*, vol. COM-30, no. 7, pp. 1679-1692, July 1982.

7. Hans Marmolin, "Subjective MSE Measures," *IEEE Trans. on Systems, Man, & Cybernetics*, vol. SMC-16, pp. 486-489, IEEE, June, 1986.

8. Sakuichi Ohtsuka, Masayuki Inoue, and Kazuhisa Watanabe, "Quality Evaluation on Pictures with Multiple Impairments Based on Visually Weighted Error," *NTT Human Interface Labs. techical report*, Kanagawa, Japan, 1987.

9. Thomas G. Stockham, Jr., "Image Processing in the Context of a Visual Model," *Proc. IEEE*, vol. 60, no. 7, pp. 828-842, July 1972.

10. Zigmantas L. Budrikis, "Visual Fidelity Criterion and Modeling," *Proc. IEEE*, vol. 60, no. 7, pp. 771-779, July 1972.

11. James L. Mannos and David J. Sakrison, "The Effects of a Visual Fidelity Criterion on the Encoding of Images," *IEEE Trans. on Info. Theory*, vol. IT-20, no. 4, pp. 525-536, July, 1974.

12. R. W. Cohen and I. Gorog, "Visual Capacity - An Image Quality Descriptor for Display Evaluation," *Proc. SID*, vol. 15, no. 2, pp. 53-62, 1974.

13. David J. Sakrison, "On the Role of the Observer and a Distortion Measure in Image Transmission," *IEEE Trans. on Comm.*, vol. COM-25, no. 11, pp. 1251-1267, Nov. 1977.

14. Charles F. Hall, "The Application of Human Visual System Models to Digital Color Image Compression," *SPIE*, vol. 594, pp. 21-28, 1985.

15. Names withheld to protect the implicated.

16. Andrew B. Watson, "Receptive fields and visual representations," *SPIE Proceedings*, vol. 1077, 1989.

17. A. B. Watson, "Efficiency of a model human image code," *J. Opt. Soc. of Am.*, vol. 4, no. 12, p. 2401, Dec. 1987.

18. D.R. Halverson, N.C. Griswold, and G.L. Wise, "A Generalized Block Truncation Coding Algorithm for Image Compression," *IEEE Trans. on Acoustics, Speech, and Signal Proc.*, vol. ASSP-32, no. 3, pp. 664 - 668, June 1984.

19. N.C. Griswold, D.R. Halverson, and G.L. Wise, "A Note on Adaptive Block Truncation Coding for Image Processing," *IEEE Trans. on Acoustics, Speech, and Signal Proc.*, vol. AQSSP-35, no. 8, pp. 1201 - 1203, Aug. 1987.

20. Edward J. Delp and O. Robert Mitchell, "Image Compression Using Block Truncation Coding," *IEEE Trans. on Comm.*, vol. Com-27, no. 9, pp. 1335-1342, Sept. 1979.

21. O. Robert Mitchell and Edward J. Delp, "Multilevel Graphics Representation Using Block Truncation Coding," *Proc. of the IEEE*, vol. Vol. 68, no. 7, pp. 868 - 873, Nov. 1982.

22. D.R. Halverson, "On the Implementation of a Block Coding Algorithm," *IEEE Trans. on Comm.*, vol. COM-30, no. 11, pp. 2482 - 2484, November 1982.

23. N. Ahmed, T. Natarajan, and K. R. Rao, "Discrete Cosine Transform," *IEEE Trans. on Computers*, pp. 90-93, Jan. 1974.

24. Wen-Hsiung Chen and William K. Pratt, "Scene Adaptive Coder," *IEEE Trans. on Comm.*, vol. COM-32, no. 3, pp. 225-232, March, 1984.

25. King N. Ngan, Kin S. Leong, and Harcharan Singh, "Cosine transform coding incorporating human visual system model," *SPIE - Visual Comm. & Image Proc.*, vol. 707, pp. 165-171, 1986.

26. Peter J. Burt and Edward H. Adelson, "The Laplacian Pyramid as a Compact Image Code," *IEEE Trans. on Comm.*, vol. 31, no. 4, pp. 532-540, April 1983.

27. A. B. Watson, "The cortex transform: Rapid computation of simulated neural images," *Computer Vision, Graphics, and Image Proc.*, pp. 311-327, 1987.

28. A. B. Watson, K. R. K. Nielsen, A. Poirson, A. Fitzhugh, A. Bilson, K. Ngugen, and A. J. Ahumada, Jr., "Use of a raster framebuffer in vision research," *Behavioral Research Methods, Instruments, & Computers*, vol. 18, no. 6, pp. 587-594, 1986.

29. A. B. Watson and D. G. Pelli, "QUEST: a Bayesian adaptive psychometric method," *Perceptual Psychophysics*, vol. 33, pp. 113-120, 1983.

30. Image No. 4.2.4-g from USC-SIPI Vol 4 - green component of full color image.

31. A. B. Watson, "Probability summation over time," *Vision Res.*, vol. 19, pp. 515-522, 1979.

32. Greg Holley, "Laplacian pyramid coding experiment notes," *unpublished*, Vision Group, NASA-Ames Research Center, 1987.

33. Curtis R. Carlson and Roger W. Cohen, *Visibility of Displayed Information*, PRRL-78-CR-34, Office of Naval Research, Arlington, VA, 1977.

Perceptual Gains for Coding of Moving Images without Visible Impairments

Bernd Girod

Massachusetts Institute of Technology Media Laboratory

20 Ames Street, Cambridge, MA 02139, U.S.A.

Invited Paper

Abstract

We discuss the significance of perceptual effects for source coding of video signals without visible impairments. A new nonlinear spatiotemporal model of human threshold vision is proposed. Linearization yields the space-time-varying w-model. The model predicts a variety of perceptual effects accurately. Maximum bit-rate savings by irrelevancy reduction according to the w-model are evaluated for natural test pictures on the basis of the Shannon Lower Bound of rate distortion theory. Under typical viewing conditions, perceptual gains due to linear effects dominate. Maximum bit-rate savings due to the nonlinear masking effects are below 0.5 bit/sample in the average. Typically $\frac{1}{3}$ of the masking gain is due to spatial masking, the rest is due to the presence of dark and bright areas in the picture, where the visibility of noise is reduced. Gains due to temporal masking are significant only in the first 100 ms after a scene cut.

1. Introduction

The bit-rate that is required for the transmission or storage of moving video signals can be reduced by source coding techniques. In order to represent a video signal by the minimum number of binary digits, two principles are utilized:

- Redundancy reduction: Properties of the information source, that are known a priori, result in redundant information that need not be transmitted.

- Irrelevancy reduction: The human observer does not perceive certain deviations of the received signal from the original.

In this paper, we will concern ourselves exclusively with irrelevancy reduction. Three questions arise in connection with irrelevancy reduction of video signals:

1. Which model reflects the properties of the human visual system adequately?

2. What bit rate savings are possible for a given fidelity criterion?

3. What is the optimum coding algorithm for a given fidelity criterion?

This contribution is concerned with questions 1 and 2. We will describe a new threshold model of human vision for irrelevancy reduction of monochrome TV signals. For further analysis, this model will be linearized to yield the so-called *w-model*. Various perceptual effects are predicted accurately by the w-model, including the spatiotemporal modulation transfer function of the human visual system, spatial masking, and temporal masking. Maximum bit-rate savings that can be obtained by exploiting perceptual effects are then calculated by means of rate distortion theory.

2. A Model of Human Threshold Vision

Fig. 1 shows the block diagram of a nonlinear model of human visual brightness perception that has been developed for the optimization of video source coding algorithms. The model is three-dimensional: it considers input luminance signals $s(x, y, t)$, that are functions of both visual angle (x, y) and time t. The output of the model is a two-valued variable that indicates whether the input signal $s(x, y, t)$ contains visible impairments. The model was developed to predict visibility thresholds measured with average or above-average test subjects under standard viewing conditions recommended for quality assessment of TV images in CCIR Rec. 500 [1]. An elaborate description of the full model and a description of the experiments to determine the model parameters can be found in [2] [3].

For source coding applications we care about visibility thresholds in the video signal domain. We thus have to include into our model the display device, that converts the video signal $s(x, y, t)$ into the screen luminance $l(x, y, t)$. Our monitor model takes into account the nonlinear "γ-characteristic" of the picture tube as well as spatiotemporal filtering effects.

The luminance signal $l(x, y, t)$ on the monitor screen is mapped into a luminance signal $l_{retina}(x, y, t)$ on the retina by the optics of the observer's eye. Our model assumes an optical point spread function that is Gaussian. Additionally, the luminance signal undergoes a coordinate transformation that depends on the current direction of the visual axis. Note that the direction of the visual axis is an additional input to the model. We can thus use the model the study the effect of arbitrary eye movements. The consequences of eye movements for coding of video sequences have been discussed comprehensively in [4].

The retina of the human eye converts the luminance signal $l_{retina}(x, y, t)$ into a train of nerve pulses that is transmitted to the central nervous system by the optical nerve. The central fovea is the most sensitive part of the retina for almost all coding impairments. We therefore model the signal processing in the human fovea, which functions as a dynamic gain control. The nonlinear spatiotemporal interactions in the model give rise to masking effects at spatiotemporal luminance discontinuities, that very accurately correspond to those observed for the human visual system. If our model is correct, inhibition and saturation effects in the human retina are the cause of both spatial and temporal masking [5]. The output of the fovea model is a hypothetical ganglion cell signal $c(x, y, t)$.

The final model stage in Fig. 1 summarizes all higher order perceptual processes, which finally might lead to a recognition of impairments in the signal $s(x, y, t)$. For such a recognition, the human visual system has to separate the impairment component $\Delta c(x, y, t)$ contained in the signal $c(x, y, t)$ from the rest of the signal. Today, we do not understand the signal analysis in the human brain that performs such separation. For additive, noiselike impairments the assumption of a perfect separation of the impairment from the remainder appears to be justified. We model this perfect separation by calculating the difference $\Delta c(x, y, t)$ between the impaired signal $c(x, y, t)$ and the signal $c^*(x, y, t)$ that would result without the impairment. When optimizing a source coding scheme, we usually have a reference video signal $s^*(x, y, t)$ available. This reference video signal is passed through the same chain of components as the impaired video signal to yield $c^*(x, y, t)$ (Fig. 1). The bubble labelled "Impairment detection" in Fig. 1 then calculates a local mean squared Δc. If this local mean squared error exceeds a fixed threshold somewhere in the image sequence, the impairment is predicted to be visible.

The visibility of $\Delta s(x, y, t)$ is highly dependent on $s^*(x, y, t)$ owing to the nonlinearity of the model. Since impairments at visibility threshold are usually small, we can linearize the model around its operating point determined by $s^*(x, y, t)$. This linearized "w-model" has an overall structure as shown in Fig. 2. Processing of the reference signal $s^*(x, y, t)$ corresponds to the processing in the nonlinear model (Fig. 1), with the weights $w_1(x, y, t)$, $w_2(x, y, t)$, and $w_3(x, y, t)$ as side results. We denote this lower branch as the *masking branch* of the w-model, since it fully captures the influence of the reference signal $s^*(x, y, t)$ on the perception of an impairment. The upper *"impairment branch"* is linear, but it is space-time varying, as the weights w_1, w_2, and w_3 depend on (x, y, t).

The w-model possesses several practical advantages over the nonlinear threshold model (Fig. 1). It is mathematically tractable. For many basic psychophysical experiments, like the measurement of the modulation transfer function, or the measurement of visibility thresholds at a step edge, an analytical calculation of the model prediction is possible. Further, the w-model can be analyzed in terms of rate distortion theory. The w-model also facilitates the interpretation of perceptual phenonema, since the influence of the impairment structure on its visibility can be separated from the influence of the large amplitude signal $s^*(x, y, t)$ as a result of the separate impairment and masking branches.

Of course, the usefulness of the w-model is determined by the range of perceptual phenomena that it can predict. In psychophysical tests we have measured perceptual thresholds for a variety of impairments $\Delta s(x, y, t)$ and reference signals $s^*(x, y, t)$. We cannot describe all these experiments in this paper; the interested reader is referred to [2]. In our experiments, we verified that the w-model predicts the following perceptual phenomena with good accuracy:

- The effect of the impairment's duration and spatial extent on its visibility

- The spatiotemporal modulation transfer function of the human visual system

- The influence of a uniform background field on the visibility of impairments

- Masking at spatial luminance edges ("spatial masking")

- Masking at temporal luminance discontinuities ("temporal masking")

3. Maximum Bit-Rate Savings by Irrelevancy Reduction

Assuming that the above model reflects the properties of human visual perception accurately, what maximum bit-rate savings are possible by exploiting perceptual effects? Rate distortion theory [6] provides the mathematical tools to treat this problem. Particularly useful is the *Shannon Lower Bound* [6]

$$R \geq h(s^*) - \max_{D \leq D_\theta} \{h(\Delta s)\}, \tag{1}$$

which states that the transmission bit-rate R has to be always larger than the entropy of the video signal source, $h(s^*)$, diminished by $\max\{h(\Delta s)\}$, the maximum value that the entropy of the impairment Δs can take on. In principle, $h(\Delta s)$ could be arbitrarily large, but not if a fidelity criterion $D \leq D_\theta$ is obeyed. The fidelity criterion postulates that the distortion D is smaller than a maximum acceptable distortion D_θ, and this limits the entropy $h(\Delta s)$. The Shannon Lower Bound allows us to split up *potential* coding gains between redundancy reduction and irrelevancy reduction. The first term of (1), the source entropy $h(s^*)$, depends only on the source statistics that are to be exploited for redundancy reduction. The second term depends only on the fidelity criterion, which is the basis for irrelevancy reduction. We call this term $\max\{h(\Delta s)\}$ *"maximum perceptual gain"*, since it captures the maximum bit-rate savings due to irrelevancy reduction.

The maximum perceptual gain $\max\{h(\Delta s)\}$ has been calculated for impairments below the visibility threshold according to the w-model. It turns out that the overall perceptual gain can be split up into several additive terms that correspond to different perceptual phenomena. Typical results are illustrated in Fig. 3. As a reference quality level we use an 8 bit/sample PCM representation of the video signal. Perceptual gains depend on the viewing distance and the sampling raster of the video signal. In this example we assume a viewing distance of six times screen height, 25 frames/sec, 575 active lines/frame, and 540 active samples/line.

The overall perceptual gain of approximately 4.5 bit/sample is due to two groups of effects. For the first group, consisting of gains due to the visibility threshold for white noise, the horizontal transfer function, the vertical transfer function, and the temporal transfer function, perceptual gains are independent of the picture contents. Note that perceptual gains corresponding to the vertical transfer function cannot be exploited with line-interlaced displays, since eye movements and interlace interact unfavorably. The second group of effects depends entirely on the picture contents, indeed the perceptual gain is defined to be zero for a blank field of medium grey. The numbers in Fig 3 are valid for typical TV material. Roughly $\frac{1}{3}$ of the masking gain is due to spatial masking, the rest is due to the presence of very dark or very bright areas in the picture with reduced visibility of noise. Gains due to temporal masking were found to be insignificant even for moving image sequences, with the exception of scene cuts. In the first 100 ms after a scene cut, significant bit-rate savings due to temporal masking are possible without visible impairments. Masking gains are discussed in detail in [3].

Some of the perceptual gains were found to be very small. This is disappointing from a system designer's point-of-view. It should be kept in mind, that these numbers are upper limits that cannot be exceeded by a practical system. The disappointingly small gains are the strongest statements of our rate distortion theoretical analysis. Almost certainly, however, there are additional masking effects in the human visual system, more complicated than those predicted by the w-model. If we succeed to incorporate these effects into a comprehensive model, it would be very valuable to investigate their significance for the encoding of video signals by the approach outlined in this contribution.

4. References

[1] "Method for the subjective assessment of the quality of television pictures", CCIR Rec. 500-2, 1982.
[2] B. Girod, "Ein Modell der menschlichen visuellen Wahrnehmung zur Irrelevanzreduktion von Fernsehluminanzsignalen," Doctoral dissertation, Universität Hannover, VDI-Fortschrittberichte, Reihe 10, Nr. 46, 1987 (in German).
[3] B. Girod, "The information theoretical significance of spatial and temporal masking in video signals," SPIE/SPSE Symposium on Electronic Imaging, Conf. on Human Vision, Visual Processing and Digital Display, Los Angeles, CA, January 1989, also submitted to Optical Engineering.
[4] B. Girod, "Eye movements and coding of video sequences," SPIE vol. 1001, Visual Communications and Image Processing '88, 398 - 405, 1988.
[5] B. Girod, "Spatial and Temporal Masking by Saturation in the Human Fovea," Proc. of the First Annual INNS Meeting, Boston, p. 497, September 1988.
[6] T. Berger, "Rate Distortion Theory," Englewood Cliffs, N.J.: Prentice-Hall, 1971.

OSA Meeting on APPLIED VISION
San Francisco, CA, July 1989

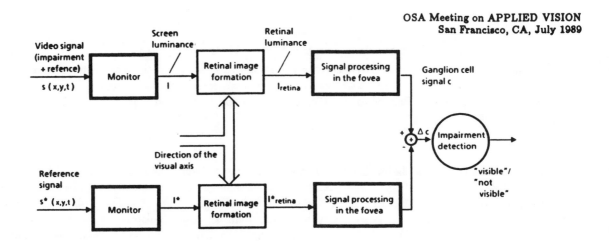

Fig. 1: Nonlinear 3-D threshold model of human brightness perception. Bold blocks indicate nonlinear system components.

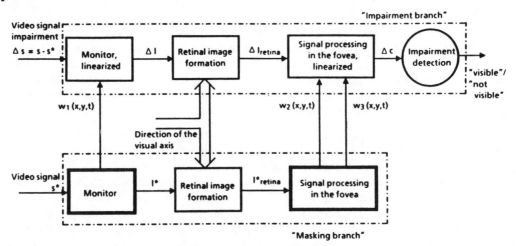

Fig. 2: Block diagram of the w-model.

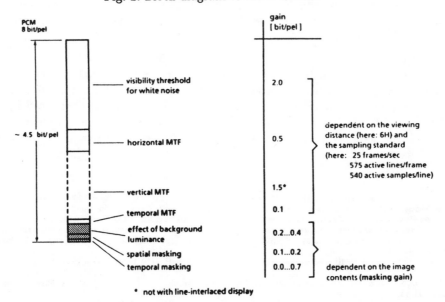

Fig. 3: Maximum perceptual gains without visible impairments according to the w-model.

IMAGE GATHERING AND DIGITAL RESTORATION:
END-TO-END OPTIMIZATION FOR VISUAL QUALITY

Friedrich O. Huck
NASA Langley Research Center, Hampton, Virginia 23665

Sarah John, Judith A. McCormick, and
Ramkumar Narayanswamy
Science and Technology Corporation, Hampton, Virginia 23666

ABSTRACT

Image gathering and digital restoration are commonly treated as separate tasks. However, it is possible to gain significant improvements in fidelity, resolution, sharpness, and clarity when these two tasks are optimized together. In this paper, we demonstrate the improvements that can be gained when (1) the design of the image-gathering system is optimized for high information density rather than for conventional image reconstruction, and (2) the digital restoration of the image accounts for the aliasing as well as the blurring and noise in image gathering and practically eliminates the degradations that occur due to the blurring and raster effects in image reconstruction.

INTRODUCTION

The goal of image gathering and digital restoration in many important applications is to acquire visual information for human interpretation. Applications typically include military reconnaissance, planetary exploration, and medical diagnosis. It is usually desired, within the critical constraints imposed on image gathering, to restore pictures of the target with the best possible fidelity (resemblance to the target), resolution (minimum discernible detail), sharpness (contrast between large areas), and clarity (absence of visible artifacts and noise). The complexity of the digital processing required to do so is usually not a severe constraint. Moreover, it is sometimes also desired to interactively enhance certain features of the target for improved resolution and contrast even at the cost of fidelity and clarity.

This goal has often been significantly and unnecessarily compromised in the past by the treatment of image gathering and digital restoration as separate tasks. Traditionally, the design of image-gathering systems and image-display devices has belonged to the realm of telephotography and television, whereas the formulation of digital image-restoration algorithms has belonged to the realm of digital signal processing. Consequently, despite the complex digital processing encountered in practice, the restored images have often failed to improve upon the visual quality obtained in a simpler and faster way by image reconstruction and interpolation. As Schreiber[1] puts it succinctly: "The effect of the characteristics of imaging components, especially cameras and display devices, on the overall performance of systems is generally ignored in the [digital processing] literature, but is actually very large." In this paper, we present a method for optimizing the end-to-end performance of image gathering and interactive digital restoration. This method accounts for the effects of the image-gathering process and practically eliminates the effects of the image-reconstruction process.

MODEL

Figure 1 illustrates a model of image gathering, digital processing, and image reconstruction. Image gathering transforms the continuous (incoherent) radiance field $L(x, y)$ into a digital signal $s(x, y)$, and image reconstruction transforms the digitally processed signal into a continuous representation $R(x, y)$. The digital signal $s(x, y)$ and the continuous representation $R(x, y)$ are defined by the expressions[2-5]

$$s(x, y) = [KL(x, y) * \tau_g(x, y)] \; \text{III} \; (x, y) + n(x, y) \qquad (1)$$

and

$$R(x, y) = [s(x, y) * K^{-1}\tau_p(x, y)] \; \text{III}' \; (x, y) * \tau_r(x, y), \qquad (2)$$

where K is the steady-state gain of the (linear) radiance-to-signal conversion, $\tau_g(x, y)$, $\tau_p(x, y)$ and $\tau_r(x, y)$ are the spatial responses of the image-gathering system, data-processing algorithm and image-display system, respectively, n(x,y) is the (additive, discrete) sensor noise, * denotes spatial convolution, and $\text{III} \; (x, y)$ and $\text{III}' \; (x, y)$ denote the image-gathering and image-display lattices, respectively, in the (x,y) rectangular coordinate system that is used as a reference for the imaging process.

We assume, for convenience, that the image-gathering lattice $\text{III} (x,y)$ has intervals of unit length. The intervals of the image-display lattice $\text{III}' (x,y) = \text{III}(\Delta x, \Delta y)$ are conventionally equal to the sampling intervals (i.e., $\Delta = 1$). However, it is possible to suppress the blurring and raster effects in image displays by letting the intervals of the display lattice be denser than those of the image-gathering lattice (i.e., $\Delta > 1$). Each traditional picture element (pel) consists then of $\Delta \times \Delta$ display elements (dels). If we allow Δ to be 4 or more, then the effects of the image-display device become negligible, and the Fourier transform of R(x,y) yields the spatial-frequency representation $\hat{R} (v, \omega)$ of the restored image

$$\hat{R}(v, \omega) = \left[\hat{L}(v, \omega)\hat{\tau}_g(v, \omega) * \; \hat{\text{III}} \; (v, \omega) + K^{-1}\hat{n}(v, \omega)\right] \hat{\tau}_p(v, \omega), \quad (3)$$

where $\hat{L} (v, \omega)$ and $\hat{n} (v, \omega)$ are the radiance-field and noise transforms, respectively, $\hat{\tau}_g (v, \omega)$ and $\hat{\tau}_p (v, \omega)$ are the spatial-frequency responses of the image-gathering system and data-processing algorithm, respectively, and v, ω are the spatial frequencies with units of cycles per sample. The function $\hat{\text{III}} (v, \omega)$ is the transform of the image-gathering lattice given by $\hat{\text{III}} (v, \omega) = \delta(v, \omega) + \sum_{\neq 0,0} \hat{\text{III}} (v, \omega),$

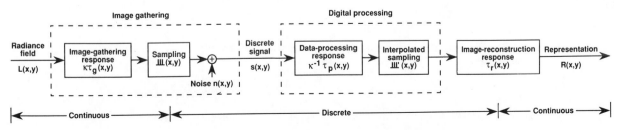

Figure 1. Model of image gathering, digital processing, and image reconstruction.

where $\underline{\underline{|||}}(v,\omega)$ accounts for the sampling sidebands. The associated sampling passband is $\hat{B} = \{(v,\omega), |\, v\,| < 1/2, |\,\omega\,| < 1/2\}$, and its area is $|\hat{B}| = 1$.

IMAGE RESTORATION

In the traditional development of image-restoration filters, the sampling required to transform the continuous-input radiance field into a digital signal is erroneously assumed to be sufficient, and the display raster required to transform the processed digital signal into the continuous-output picture is ignored. Hence, the Wiener restoration filter for the digital processing algorithm $\hat{\tau}_p(v,\omega)$ is (wrongly) given in the prevailing literature by

$$\hat{W}(v,\omega;\gamma) = \frac{\hat{\Phi}'_L(v,\omega)\hat{\tau}_g^*(v,\omega)}{\hat{\Phi}'_L(v,\omega)\,|\,\hat{\tau}_g(v,\omega)\,|^2 + \gamma(K\sigma_L/\sigma_N)^{-2}} \qquad (4)$$

where $\hat{\Phi}'_L(v,\omega) = \sigma_L^2\hat{\Phi}'_L(v,\omega)$ is the Wiener spectrum of the incident radiance field $L(x,y)$, σ_L^2 is the corresponding variance, and $\gamma = 1$. The parameter γ is often added to control the visual quality of the restored image. Other ad hoc modifications are also used.

However, if the continuous-to-discrete transformation in image gathering is correctly accounted for and the blurring and raster effects in image-reconstruction are practically eliminated by digital interpolation prior to reconstruction, then the Wiener restoration filter for $\tau_p(v,\omega)$ is (correctly) given by[2-5]

$$\hat{\Psi}(v,\omega) = \frac{\hat{\Phi}'_L(v,\omega)\hat{\tau}_g^*(v,\omega)}{\hat{\Phi}'_L(v,\omega)\,|\,\hat{\tau}_g(v,\omega)\,|^2 * \underline{\underline{|||}}(v,\omega) + (K\sigma_L/\sigma_N)^{-2}}. \qquad (5)$$

Fales et al.[4] extended this formulation to account for the discrete-to-continuous transformation in display devices so that the density of the display lattice could be relaxed (usually to $\Delta = 2$ or 3) without significant display degradations. They also demonstrated that the pictures restored with the Wiener filter $\hat{\Psi}(v,\omega)$ given by Eq. (5) are significantly superior in fidelity, resolution, and sharpness to those restored by the conventional Wiener filter $\widehat{W}(v,\omega)$ given by Eq. (4). But these fidelity-maximized images also exhibit some visual defects such as ringing, aliasing artifacts, and noise. Consequently, McCormick et al.[5] added some ad hoc features to the Wiener filter that allow an operator to interactively control the tradeoff between the enhancement of high-frequency contrast and the suppression of visual defects. The resultant interactive image-restoration filter, referred to as Wiener-Gaussian enhancement filter (WIGE), is given by[5]

$$\hat{\Psi}_i(v,\omega) = \hat{\Psi}(v,\omega)\left\{\exp\left[-2(\pi\sigma_i\rho)^2\right] + \zeta(2\pi\rho)^2\exp\left[-2(\pi\sigma_e\rho)^2\right]\right\}, \qquad (6)$$

where ζ is the enhancement parameter that controls the relative amount of the synthetic-high filtered image component in the restored image. The standard deviation σ_i controls the smoothing of the low-pass filtered image component, and the standard deviation σ_e controls the smoothing associated with edge enhancement. The synthetic-high edge enhancement was introduced to reduce the ringing (Gibbs phenomenon) that is associated with the steep roll off in the Wiener filter.

IMAGE GATHERING

Image-gathering systems have traditionally been designed to maximize the visual quality of reconstructed images. However, the design goal should now be to minimize the constraint that image gathering imposes on our ability to restore and enhance pictures (restorability). It is, therefore, intuitively attractive to treat image gathering like a communication channel. One of Shannon's theorems[6] (the 21st) states that the communication-channel design that maximizes the information density of the acquired (sufficiently sampled) data also maximizes the fidelity of optimally restored representations of the continuous-input source. Furthermore, Linfoot[7] anticipated that informationally optimized image gathering offers the opportunity for extracting optimally the widest range of spatial features. Based on extensive experience with optical systems, Linfoot related information to visual quality as follows: "An optical system can properly be said to be of high quality only if the amount of information contained in its image approaches the maximum possible . . ., and it is an agreeable consequence . . . that those which are efficient according to this criterion also form images which are sharp and clear in the usual sense of the words. "

However, neither Shannon nor Linfoot needed to account for the insufficient sampling (i.e., aliasing) that constrains not only digital image gathering but also natural vision.[8] Huck et al.[2,3] found that the combined process of image gathering and optimal processing can be treated as a communication channel if (and only if) the image-gathering degradations, including aliasing, are correctly accounted for. They also found that the robustness of the optimal restorations to uncertainties in the incident radiance-field statistics increases with increasing information density.

The spectral information density $\hat{h}(v,\omega)$ of the discrete signal $s(x,y)$ is given by[2,3]

$$\hat{h}(v,\omega) = log_2\left[1 + \frac{\hat{\Phi}'_L(v,\omega)\,|\,\hat{\tau}_g(v,\omega)\,|^2}{\hat{\Phi}'_L(v,\omega)\,|\,\hat{\tau}_g(v,\omega)\,|^2 * \underline{\underline{|||}}_{\neq0,0}(v,\omega) + (K\sigma_L/\sigma_N)^{-2}}\right] \qquad (7)$$

The corresponding (total) information density h is[2,3]

$$h = \frac{1}{2}\int\int_{\hat{B}}\hat{h}(v,\omega)dvd\omega. \qquad (8)$$

The information density h is independent of processing, provided that the spatial-frequency response $\hat{\tau}_p(v,\omega)$ of the processing, regardless of its shape, extends to or beyond the sampling passband $\hat{B}$ and the image reconstruction does not introduce noise (e.g., film

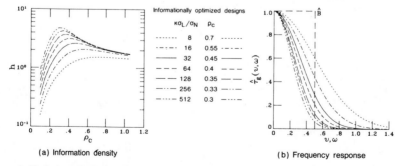

Figure 2. Variation of information density with image-gathering system design. The informationally optimized design is the design for which the image-gathering response designated by ρ_c is selected to maximize the information density h for a given SNR $K\sigma_L/\sigma$N.

granularity). Although information is not itself a direct measure of visual quality, it nevertheless proves to be a sensitive measure of restorability. It is possible, as Huck et al. [2,3] have shown, to relate the spectral information density $\hat{h}(v,\omega)$ given by Eq. (7) to the Wiener filter $\hat{\Psi}(v,\omega)$ given by Eq. (5) as follows:

$$\hat{\Psi}(v,\omega) = \frac{1}{\hat{\tau}_g(v,\omega)}\left[1 - 2^{-\hat{h}(v,\omega)}\right] \qquad (9)$$

As can be seen, the restorability of images is solely limited by the term $2^{-\hat{h}(v,\omega)}$.

Figure 2 illustrates the dependence of information density on the image-gathering response for several SNR's, ranging from the low value of 8 to the high value of 512. These results demonstrate that the informationally optimized tradeoff between aliasing and blurring, and hence the relationship between image-gathering response and sampling passband, depends on the SNR. This is intuitively appealing. In one extreme, when the SNR is very low, the enhancement of contrast is constrained mostly by noise. Hence, the image-gathering response should then be selected to reduce blurring at the cost of aliasing. In the other extreme, when the SNR is very high, the enhancement of contrast is constrained mostly by aliasing. Hence, the image-gathering response should then be selected to reduce aliasing at the cost of blurring.

END-TO-END PERFORMANCE

Figure 3 compares image gathering and restoration responses of the traditional and optimal methods. The response of the traditional (incorrect) Wiener filter $\widehat{W}(v,\omega)$ is circularly symmetric and extends far beyond the sampling passband. Textbooks are generally vague about the digital implementation of this filter. One common method, as shown here, is to truncate the response of the Wiener filter by the square sampling passband and then to Fourier transform the discrete representation of this truncated response into the spatial domain and there convolve it with the image data. The response of the (correct) Wiener filter $\hat{\Psi}(v,\omega)$ looses the circular symmetry and essentially limits the restoration to the spatial-frequency components that are located within the sampling passband. Furthermore, the Fourier transform of the discrete representation of this filter is implemented with a display lattice that is $\Delta = 4$ times denser than the sampling lattice of the image-gathering system.

Figure 4 illustrates the images produced by the two methods. The SNR was chosen to be high so that the noise would not ob-

scure the degradations caused by the image gathering and reconstruction processes, and the targets were chosen for their sensitivity to these degradations. The bar width of the targets varies from two times the sampling interval ($\mu = 2$) to half the sampling interval ($\mu = 1/2$) or less. That is, the fundamental spatial frequencies for $\mu > 1$ fall inside the sampling passband and the other frequencies fall outside this passband. If there were no image gathering and reconstruction degradations other than the sampling passband constraint, then it should ideally be possible to restore the frequency components inside the sampling passband.

As can be seen, the optimal method significantly improves upon the fidelity, resolution, sharpness, and clarity produced by the traditional method. The visual quality of the latter is degraded by familiar flaws. Bars that are larger than the sampling interval ($\mu > 1$) appear with random changes in dimensions and grey levels and with stair-step distortions ("jaggies"), while bars that are smaller than the sampling interval ($\mu < 1$) appear as moiré patterns. By contrast, the optimal method practically eliminates these flaws and correctly reproduces frequency components almost out to the sampling passband limit.

REFERENCES

1. W. F. Schreiber, *Fundamentals of Electronic Imaging Systems.* (Springer-Verlag, Berlin, 1986).
2. F. O. Huck, C. L. Fales, N. Halyo, R. W. Samms, and K. Stacy, "Image gathering and processing: Information and fidelity, "J. Opt. Soc. Am. A2, 1644-1666 (1985).
3. F. O. Huck, C. L. Fales, J. A. McCormick, and S. K. Park, "Image-gathering system design for information and fidelity, "J. Opt. Soc. Am. A5, 285-299 (1988).
4. C. L. Fales, F. O. Huck, J. A. McCormick, and S. K. Park, "Wiener restoration of sampled image data: End-to-end analysis," J. Opt. Soc. Am. A5, 300-314 (1988).
5. J. A. McCormick, R. Alter-Gartenberg, F. O. Huck, "Image gathering and restoration: information and visual quality, "J. Opt. Soc. Am., accepted.
6. C. E. Shannon, "A mathematical theory of communication, "Bell Syst. Tech. J. 27, 379-423, and 28, 623-656 (1948); C. E. Shannon and W. Weaver, The Mathematical Theory of Communication (U. Illinois Press, Urbanna, 1964).
7. E. H. Linfoot, "Transmission factors and optical design, "J. Opt. Soc. Am. 46, 740-752 (1956).
8. H. B. Barlow, "Critical limiting factors in the design of the eye and visual cortex, "Proc. R. Soc. London Ser. B212, 1 (1981).

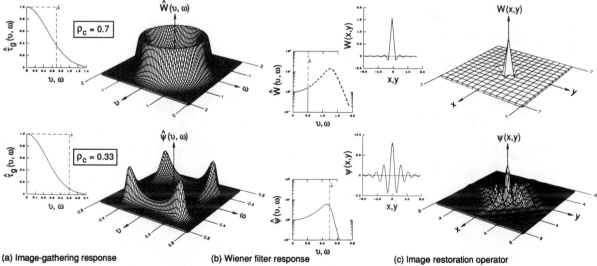

(a) Image-gathering response   (b) Wiener filter response   (c) Image restoration operator

Figure 3. Comparison of the image gathering and digital restoration responses for the traditional (upper row) and optimal (lower row) methods. The SNR $K\sigma_L/\sigma_N = 256$.

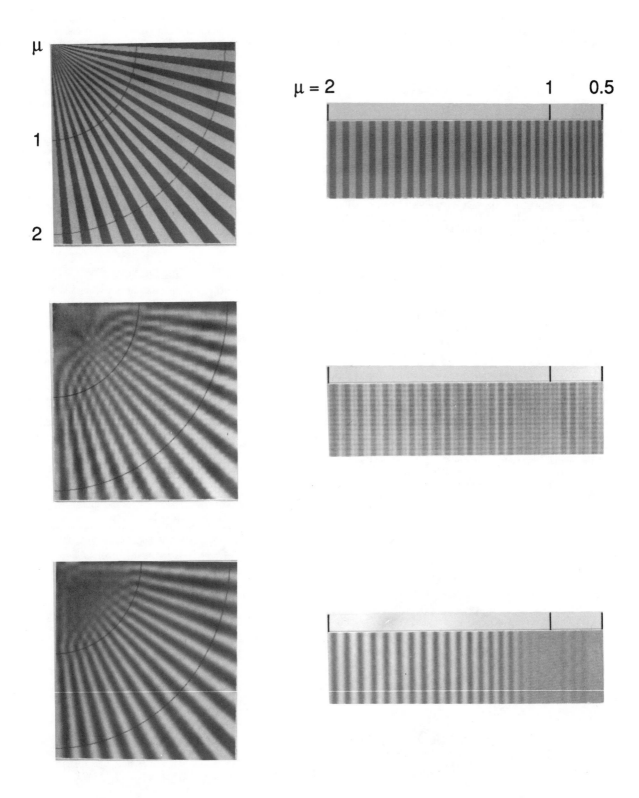

Figure 4. Comparison of the visual quality produced by the two different methods for image gathering and digital restoration. Upper row: Targets. Middle row: Traditional method. Lower row: Optimal method with WIGE restoration. The SNR $K\sigma_L/\sigma_N = 256$.

THURSDAY, JULY 13, 1989

1:30 PM–3:15 PM

ThB1–ThB5

READING AND DISPLAY LEGIBILITY

Curt R. Carlson, SRI David Sarnoff Research Center, *Presider*

READING: EFFECTS OF CONTRAST AND SPATIAL FREQUENCY[1]

Gordon E. Legge

University of Minnesota, Minneapolis, MN 55455.

Introduction

Reading is a complex everyday task. Successful reading requires high-speed visual information processing. For several years, my colleagues and I have been studying visual factors in reading with two major goals in mind: to understand the roles played by sensory mechanisms in reading and to understand how visual impairment affects reading. In a typical study, we examine the effect of an important text variable (e.g. contrast) on reading by people with normal vision. Taking the normal data as a bench mark, we try to explain abnormalities in the performance of low-vision subjects.

In this talk, I will discuss the effects of contrast and spatial-frequency filtering on reading. After reviewing the empirical findings, I will comment on the likely sensory basis for the effects and point to the relevance of the findings for low vision.

Methods

Our primary measure of reading performance is *reading speed* in words/minute. We use two procedures to evaluate a subject's maximum reading speed. In the first, a line of text drifts smoothly across the screen of a TV monitor. The subject reads the text aloud as it drifts by. If no errors are made, the drift rate is increased on the next trial. This process continues until the subject begins to make errors. Because there is a sharp transition from error-free reading to a drift rate with many errors, this technique yields a reproducible and well-defined measure of reading performance. In our second procedure, a sentence appears on the TV screen in a static display for a timed period. The subject reads through the sentence as rapidly as possible. If the subject completes the sentence, the exposure duration is reduced. Eventually, the subject cannot complete the sentence and reading speed is computed as the number of words read divided by the exposure time.

Reading rates obtained with these two procedures are highly correlated across subjects. Normal subjects can read the static text slightly faster than the drifting text but the reverse is true, on average, for subjects with low vision.

---

[1]*Supported by NIH Grant EY02934.*

How Does Luminance Contrast Affect Reading Speed?

Normal vision is quite tolerant to contrast loss (Legge, Rubin & Luebker, 1987). A tenfold reduction from maximum contrast[2] results in only a slight decrease in reading speed (less than a factor of two.) A further reduction in text contrast results in a much sharper decline in reading rate. Curves of reading speed as a function of contrast for different character sizes superimpose when text contrast is normalized by threshold contrasts for the letters.

This pattern of results can be related to psychophysical models of contrast coding. Such models typically contain a contrast-transfer function that relates visual response to stimulus contrast. The transfer function is compressive at high contrasts and can account for the tolerance of reading to contrast change. The transfer function is accelerating at low contrast and can account for the strong dependence of reading on contrast at low levels.

High contrast is critical for many people with low vision. Their reading speeds decline with any reduction from maximum contrast. Rubin & Legge (1989) showed that a subgroup of subjects with low vision--those with cloudy ocular media but no retinal involvement--can be characterized as "contrast attenuators." Their performance is normal, apart from a scaling of text contrast that is equal to their decrease from normal in letter-contrast sensitivity.

How Does Color Contrast Affect Reading Speed?

We have measured reading speed for text conveyed by color contrast (e.g. red letters on a green background) rather than the more usual luminance contrast (Parish, Legge & Luebker, 1989). Colors were matched for luminance by flicker photometry. Color contrast was varied by mixing different proportions of red and green (or yellow and blue) in the text and background. Color contrast was defined as the luminance contrast of the component colors making up the stimulus.

When color contrast is high, reading rates match the highest values found for luminance contrast. As with luminance, reading performance first declines slowly with reduction in color contrast, and then more rapidly at low levels. Reading rates for luminance and color contrast can be compared if both are expressed as multiples of threshold contrast. When this is done, curves of reading speed vs. contrast superimpose for luminance and color. This finding holds for both normal and low vision.

These results show that reading can rely equally well on information conveyed by luminance or color contrast. The mechanisms coding this information must be quite

---

[2]Text contrast is defined as (C-B)/(C+B) where $\underline{C}$ and $\underline{B}$ are the luminances of the characters and the background.

similar and possibly share a common neural pathway.

How Does Character Size Affect Reading Speed?

Reading rates are highest on a plateau extending from about $.3^o$ to 2° (Legge *et al.*, 1985a). There is a sharp decline in reading speed for smaller characters, and a more gradual decline for larger ones. Of relevance to low vision, it is important to note that normal vision can sustain functionally useful reading rates for enormous characters (e.g. 70 words/minute for 24° characters).

We have studied character-size effects in low vision (Legge, *et al.*, 1985b). Prescription of appropriate magnification in reading aids relies on identification of the range of character sizes for which reading is fastest. For a subgroup of low-vision subjects--those with large central scotomas--reading performance benefits from ever-increasing character size.

It is possible to define a *contrast sensitivity function* (CSF) for reading. This is done by finding the contrast required to produce a threshold reading rate (say 35 words/minute) at each of many character sizes. These threshold contrasts (or their reciprocals) are plotted as a function of the fundamental frequency[3] of the characters. Such plots are qualitatively similar in shape to sine-wave grating CSFs. When appropriate stimulus conditions are compared, there is a striking quantitative similarity as well.

These results suggest that character size effects in reading are related to differences in contrast sensitivity across spatial frequency. Apparently, mechanisms that limit spatiotemporal contrast sensitivity play a role in reading.

How do Low- and High-Pass Spatial Filtering Affect Reading Speed?

We measured reading rates for text that was low-pass spatial-frequency filtered (Legge *et al.*, 1985a). Reading speed was quite tolerant to bandwidth reduction (blur). Performance was unaffected except when the bandwidth extended less than one octave above the fundamental frequency of the characters. The *critical bandwidth* of one octave was constant across a wide range of character sizes. The fact that rapid reading requires just one octave in spatial frequency implies that only one spatial-frequency channel is required for reading.

While blurry letters may be highly legible, what about high-pass-filtered text? A traditional view of reading holds that important information is conveyed by coarse word-shape information. This information might be carried by low spatial frequencies, below those carrying letter information (Morris, 1988). If word-shape information plays a role in the dynamics of reading, performance should be impaired if

---

[3]The *fundamental frequency* of characters in text is equal to the reciprocal of character size in degrees.

text is high-pass filtered. Recent measurements in my lab indicate that this is not the case. There is little difference in reading speed for unfiltered and high-pass filtered text.

Conclusions

Psychophysical techniques can be used to study the role of vision in reading. The dependence of reading speed on important text variables such as contrast, character size and spatial-frequency content can be related to known characteristics of sensory mechanisms. An understanding of the role of vision in normal reading enables us to understand reading deficits in people with low vision.

References

Legge G. E., Pelli D. G., Rubin G. S. and Schleske M. M. (1985a) Psychophysics of reading. I. Normal vision. *Vision Res.* 25, 239-252.

Legge G. E., Rubin G. S. and Luebker A. (1987) Psychophysics of reading. V. The role of contrast in normal vision. *Vision Res.* 27, 1165-1171.

Legge G. E., Rubin G. S., Pelli D. G. and Schleske M. M. (1985b) Psychophysics of reading. II. Low vision. *Vision Res.* 25, 253-266.

Morris R. (1988) Image processing aspects of type. *Proc. EP88 Conf. on Text Processing*, Cambridge University Press.

Parish D.H., Legge G.E. and Luebker A. (1989) Reading and color contrast. Annual Meeting of the Association for Research in Vision and Ophthalmology, Sarasota.

Rubin G. S. and Legge G. E. (1989) Psychophysics of reading. VI. The role of contrast in low vision. *Vision Res.*, 29, 79-91.

WHY WAS READING SLOWER FROM CRT DISPLAYS THAN FROM PAPER?

John D. Gould
IBM Research Center
P O Box 704
Yorktown Heights, New York 10598

Computer displays allow people to do many things that they could not do without them. Reading from these displays is a main behavioral activity that cuts across most uses of computers. Experiments have shown that people read more slowly from CRT displays than from paper, sometimes 20-30% slower (e.g., Gould & Grischkowsky, 1984). From 1983-1985, we conducted fifteen experiments and several more analyses in an effort to understand the cause of this reading difference. Initially, each experiment isolated one variable and studied whether that variable explained the difference (Gould, Alfaro, Barnes, Finn, Grishkowsky, and Minuto, 1987). Typically, experimental participants would proofread several pages of text in good quality fonts on Paper, and then (or before) proofread similar material from a computer-controlled CRT display. Proofreading time and accuracy was recorded, and the personal feelings of participants were also noted afterwards.

No one variable studied (e.g., experience in using CRT displays; display orientation; character size, font, polarity, aspect ratio, contrast, different displays) explained why this reading speed difference occurred. There was not speed/accuracy tradeoff.

We then turned from looking for an explanation of the reading speed difference in terms of individual display variables to searching for conditions in which people could read as fast from CRT displays as from Paper (Gould, Alfaro, Finn, Haupt, and Minuto, 1987). The approach was to make CRT displays look similar to paper. In several experiments we used higher resolution displays and anti-aliased characters of paper-like fonts (Sholtz, 1984) produced by the YODA system developed by Dave Bantz and Satish Gupta at Yorktown. As shown below, we found that people did read as fast from CRT displays as they did from good quality paper fonts.

In the following three experiments participants proofread for misspelled words (about one every 150 words). The format, font, color, polarity, and size of the characters on Paper were the same as those on the screen. That is, if a transparency made of a page of paper was placed on the screen, it matched the screen layout.

Table 1 shows the results from an experiment in which 18 volunteers proofread six 5-page articles. They proofread three articles on paper (each in a different font) and three articles on the CRT (each in a different font). Articles were balanced across the six experimental conditions. Participants proofread significantly faster from Paper than from the CRT (Means= 220 and 209 words per minute (wpm), $F(1,17)=6.72$; $p<.05$). The 95% confidence interval for this mean difference was 11+-9 wpm. This means that, at the 95% confidence level, the true difference between these two means is between 2 and 20 wpm in favor of Paper. This 5% difference was the smallest we had so far found between Paper and CRT displays.

Table 1. Proofreading rate (wpm) and accuracy (%hits).

	Letter-Gothic		Press		Universe		MEANS	
Paper	217	(74)	226	(68)	219	(64)	220	(69)
CRT	202	(75)	212	(67)	213	(67)	209	(70)
MEANS	209	(74)	219	(67)	216	(66)	215	(69)

In the above experiment a color Mitsubishi display (640 x 480 addressability) was used. This display gave poor contrast when the characters were shown in monochrome. Table 2 shows results from a second experiment in which 16 different participants proofread four 1000-word articles, two from Paper and two from an IBM monochrome 5080 display (1000 x 1000 addressability). Participants proofread at about the same rate from Paper and from the CRT (Means= 201 and 196 words per minute (wpm), $F(1,15)=1.31$; $p>.10$). The 95% confidence interval for this mean difference was 5 +- 11 wpm. This means that, at the 95% confidence level, the true difference between these two means is between 16 wpm in favor of Paper and 6 wpm in favor of CRT. There was no difference in accuracy of proofreading (Means=78% and 73% hits; $F(1,15)<1.0$).

Table 2. Proofreading rate (wpm) and accuracy (%hits).

	22-line pages		28-line pages		MEANS	
Paper	198	(77)	205	(79)	201	(78)
CRT	190	(73)	203	(73)	196	(73)
MEANS	194	(75)	204	(76)	199	(76)

The 5080 display regeneration rate was 50 hertz, and some participants reported it flickered. We obtained an experimental IBM 5080 monochrome display which regenerated at 60 hertz. Twelve different participants proofread four 1000-word articles, two from Paper and two from the CRT display. Participants proofread at about the same rate from Paper and from the CRT (Table 3; Means= 206 and 204 words per minute (wpm), $F(1,15)<1.0$ The 95% confidence interval for this mean difference was 2 +- 14 wpm. This means that, at the 95% confidence level, the true difference between these two means is between 16 wpm in favor of Paper and 12 wpm in favor of CRT. There was no difference in accuracy of proofreading (Means=81% and 79% hits on Paper and CRT, respectively; $F(1,11)<1.0$). There were few false-positives (0.67 per article).

Table 3. Proofreading rate (wpm) and accuracy (%hits).

	22-line pages	28-line pages	MEANS
Paper	200 (82)	212 (79)	206 (81)
CRT	204 (81)	204 (77)	204 (79)
MEANS	202 (81)	208 (78)	205 (80)

Having found display conditions from which experimental participants read as fast as from Paper, we then tried to figure out what factors contributed most to this result. Display resolution appeared to be one important variable. Figure 1 (from Gould, Alfaro, Finn, Haupt, and Minuto, 1987) shows that proofreading speed dramatically improved with increases in display resolution. Figure 1 plots the within-participant ratio of reading speed on CRT display to that on paper. Display resolution was not studied as an independent variable; rather the data are from the various experiments we did with different displays.

In general, the evidence suggests that image quality variables associated with the display and the characters themselves account for the reading speed difference. The evidence rules out an explanation associated with people themselves (e.g., they lack familiarity or experience with CRTs), or an explanation based upon something inherently wrong with CRT technology.

References

Gould, J. D. & Grischkowsky, N. (1984). Doing the Same Work With Hardcopy And With Cathode Ray Tube (CRT) Computer Terminals. Human Factors, 26, 323-338.

Gould, J. D., Alfaro, L., Barnes, V., Finn, R., Grishkowsky, N. (1987). Reading is slower from CRT displays than from Paper: Attempts to Isolate a Single-Variable Explanation. Human Factors, 29(3), 269-299.

Gould, J. D., Alfaro, Finn, R., Haupt, B., and Minuto, A. (1987). Reading from CRT Displays Can Be as Fast as Reading From Paper Human Factors, 29(5), 497-517.

Sholtz, P. N. Making high-quality colored images on raster displays. IBM Research Report, RC-9632, 1984.

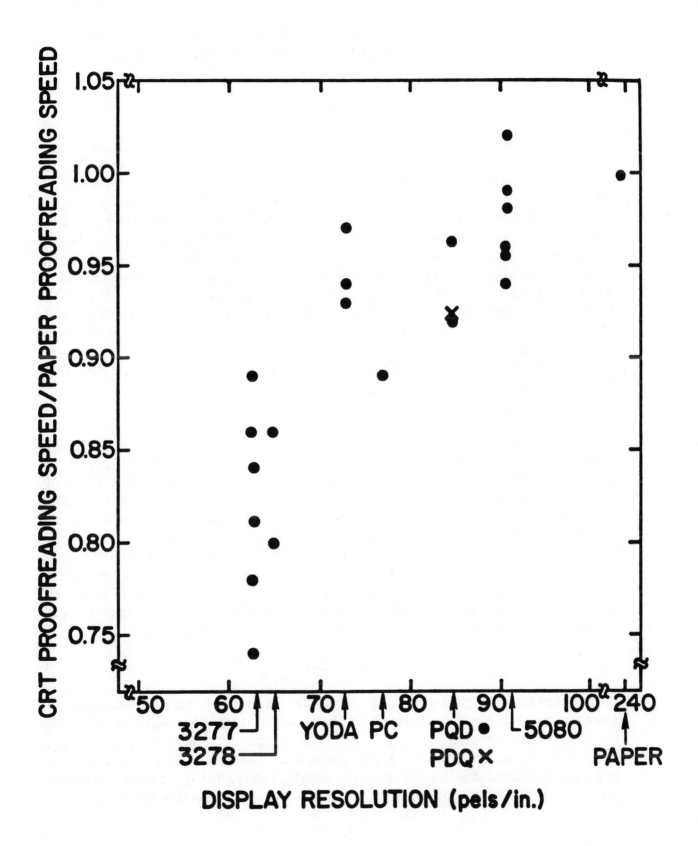

Effects of character size and chromatic contrast on reading performance.

Kenneth Knoblauch and Aries Arditi, Vision Research Laboratory,
The Lighthouse, 111 East 59th Street, New York, NY 10022

Introduction

Recent work on the psychophysics of reading has concentrated primarily on achromatic (or luminance) contrast factors (Legge *et al*, 1985, 1986, 1987). In general, reading is fastest with text of high contrast defined by luminance contours. In the real world, however, chromatic as well as luminance edges abound. While several studies have examined the interaction of chromatic and luminance contrasts on stimuli near threshold (e.g., Mullen, 1987; Knoblauch *et al*, 1984; Jameson, 1985; Switkes *et al*, 1988), only a few have investigated suprathreshold levels. Legge *et al*, (1986) demonstrated that for observers with normal vision, the luminance contrast and not the color of text on a dark background determines reading rate. This finding is consistent with threshold studies that show that the color of a grating does not affect luminance contrast sensitivity either (Nelson and Halberg, 1979), provided that one avoids frequencies near the diffraction limit (Pokorny *et al*, 1968; Van Nes and Bouman, 1967).

Using stimuli of one color on a background of a different color, Lippert (1986) found that legibility as measured with a reaction time task could be predicted with a color difference metric based on a modification of CIE 1976 (L*,u*,v*) uniform color space. He presented a nomogram for assessing the legibility of character/background pairs of different contrasts for characters in the size range of 0.36-0.55 degrees of visual angle. These character sizes might be too small to be legible for the chromatic systems (see Discussion). Thus, his results might be accounted for on the basis of luminance contrast alone. We examined reading performance for a series of color contrast conditions with a range of character widths in order to evaluate any interaction between character size and color contrast on reading performance.

Method

The contrast of contours that define text against a background can be characterized ideally as varying in luminance, (L), chromaticity, (C) or luminance and chromaticity (L + C). Unfortunately, there is no generally agreed upon metric for specifying chromatic contrast. Thus, in order to evaluate the contribution of chromatic contrast to legibility, indirect methods must be used. We compared performance for reading text defined by L or

L + C contrasts but matched for L contrast. Any difference in performance between these two conditions must be due to the presence of chromatic contrast.

The background for all conditions was set at 48 cd/m² at chromaticity coordinates (0.31, 0.35). An 80 character line of right justified text was drifted at fixed velocity through a window of width 6 characters. Text velocity was increased until the observer began to make a few errors. The minimum velocity for which the observer made at least one error adjusted for the percentage correct was taken as a measure of the maximum reading rate. Character sizes spanning the range 0.17-6.0 degrees were tested.

Results

As L contrast was decreased from 0.96 to 0.12, reading rate decreased by on average 25%. This finding is consistent with the results of Legge el al (1987) who have shown normal reading rate to be somewhat immune to contrast decrement over an extensive range of contrasts. For the range of character widths examined here, reading performance varied by only about a factor of 2, peaking between 0.3 and 1.0 degrees. When L contrast was decreased to 0.02, reading rate fell off most sharply for the smallest and largest character sizes. This resulted in a shift of the peak of the reading function to between 1 and 4 degrees.

Introduction of C contrast to text of 0.12 L contrast produced no systematic effect on reading rate. One observer showed increases in reading rate of up to 25% for character widths above 1.0 degrees, but performance of a second observer deteriorated slightly at the largest character width. If reading rate at 0.96 L contrast is at a ceiling, then it is unlikely that one could see a large increment in performance at 0.12 L contrast, since there was only a small decrement due to the L component, alone.

Reading rates for near equiluminant contrasts (L + C for which L contrast was 0.01 or lower) did not exceed those of high L contrast text at any character size. In general, reading rate for near equiluminant text did exceed that of 0.02 L contrast text. Interestingly, the largest increase in reading rate for this condition was shown for small character widths, being at least a factor of 10 for character widths less than 1 degree in one observer.

Discussion

The effect of chromatic contrast on reading performance depends on the level of luminance contrast present and on character size. At high levels of luminance contrast, reading rate is already high, and, at least for normal observers, the introduction of chromatic contrast provided little additional increment in legibility. When luminance contrast levels were sufficiently lowered, however, chromatic contrast did support fairly high reading rates.

Recent measurements of the contrast sensitivity function for chromatic gratings place the high frequency cutoff at between 10 and 12 cycle/degree (Mullen, 1985). These values set the minimum resolvable gap width of a letter for the chromatic systems at between 0.04 and 0.05 degrees. If we assume letters the dimension of standard optotypes used in acuity testing, then the minimum size for resolving chromatic spatial variation of letters is in the range 0.21-0.25 degrees.

A different approach is to consider the minimum bandwidth for reading. Based on performance with low-pass filtered text, Legge *et al* (1985) have shown that a minimum of 2 cycles/character are needed to sustain reading performance. Following the convention used by Legge *et al* (1985) to specify their filtered text, the point at which the chromatic contrast sensitivity function falls to $1/e$ of its peak is at about 3 cycles/degree (based on Mullen, 1985). This criterion implies that text with characters smaller than 0.67 degrees should be illegible to the systems in our eye that follow chromatic spatial variation. A possible explanation for the high reading rates that we saw at small character sizes is the introduction of luminance edges through uncorrected chromatic aberrations in the human eye.

References

Jameson, D. (1985) Opponent-colours theory in the light of physiological findings. *Central and Peripheral Mechanisms of Colour Vision.* D. Ottoson and S. Zeki (Eds.) London: MacMillan Press Ltd.

Knoblauch,K. Jameson, D., Hurvich, L. M., and Geller, A. (1984) Chromatic factors in threshold and suprathreshold spatial vision. *Investigative Ophthalmology and Visual Science (Suppl.)*, 25, 232.

Legge, G. E., Pelli, D. G., Rubin, G. S. and Schleske, M. M. (1985) Psychophysics of reading. I. Normal vision. *Vision Research*, 25, 239-252.

Legge, G. E. and Rubin, G. S. (1986) Psychophysics of reading. IV. Wavelength effects in normal and low vision. *Journal of the Optical Society of America*, -A3, 40-51.

Legge, G. E., Rubin, G. S. and Luebker, A. (1987) Psychophysics of reading. V. The role of contrast in normal vision. *Vision Research*, 27, 1165-1177.

Lippert, T. M. (1986) *Society for Information Display 86 Digest*, 86-89.

Mullen, K. T. (1985) The contrast sensitivity of human colour vision to red-green and blue-yellow chromatic gratings. *Journal of Physiology*, 359, 381-400.

Mullen, K. T. (1987) Spatial influences on chromatic opponent contributions to pattern detection. *Vision Research*, 27, 829-839.

Nelson, M. A. and Halberg, R. L. (1979) Visual contrast sensitivity obtained with colored and achromatic gratings. *Human Factors*, 21(2), 225-228.

Pokorny, J., Graham, C. H. and Lanson, R. N. (1968) Effect of wavelength on foveal grating acuity. *Journal of the Optical Society of America*, 58, 1410-1414.

Switkes, E., Bradley, A. and De Valois, K. K. (1988) Contrast dependence and mechanisms of masking interactions among chromatic and luminance gratings. *Journal of the Optical Society of Amer ica*, A5, 1149-1162.

Van Nes, F. L. and Bouman, M. A. (1967) Spatial modulation transfer in the human eye. *Journal of the Optical Society of America*, 57, 401-406.

Acknowledgements

This work was supported by NIH grants EY07747 and AG06551 and NASA-AMES grant NCC2-541.

Reading and Contrast Adaptation

Denis G. Pelli
Institute for Sensory Research
Syracuse University
Syracuse, NY 13244-5290

Lunn and Banks (1986), studying the reasons for "visual fatigue" among users of video text displays, found that reading from a video display caused a ten-fold elevation of the contrast threshold for a sinusoidal grating with the same spatial frequency as the lines of text. They suggested that this may contribute to "visual fatigue," possibly by affecting accommodation, but they did not explain why printed text would not have the same effect.

The printed page is nearly always black on white. Until recently, video text displays (including that of Lunn and Banks) have been predominantly white on black. Could contrast polarity be the culprit?

Methods

We presented text and sinusoidal gratings on a Macintosh II display (66 Hz frame rate) viewed from 50 cm. The white areas of the text were at 202 cd/m^2 and the black areas were 1 cd/m^2. The mean luminance of the sinusoidal gratings was 101 cd/m^2. The text, either white on black, or black on white, was 12 point Bookman font, with 15 point leading. The sinusoidal grating was horizontal with a spatial frequency of 3 cycles/degree, equal to the line frequency of the text. The adapting pattern (text or grating) was presented for 1 minute, then test trials were presented for 10 seconds, after which the subject re-adapted to the text or grating for another 10 seconds, and so on. 82% correct two-alternative forced threshold contrasts were determined using the IDEAL staircase (Pelli, 1987).

Results

White on black text, unlike black on white text, has a high contrast sinusoidal component at the line frequency, 58% in our experiments. Reading white-on-black text for 1 minute produces a noticeable afterimage and elevates threshold contrast at the line frequency by 0.8 log units. Both effects are mimicked by steadily fixating a 58%-contrast sinusoidal grating with the same line frequency. Free viewing the same grating does not produce an after image, and results in a much smaller threshold elevation: 0.4 log units.

Lunn & Banks(1986) *Human Factors* **28**:457-464.
Supported by EY04432.

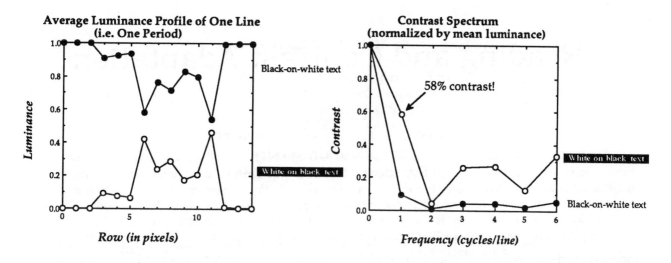

Average Luminance Profile of One Line (i.e. One Period)

Black-on-white text

White on black text

Row (in pixels)

Contrast Spectrum (normalized by mean luminance)

58% contrast!

White on black text

Black-on-white text

Frequency (cycles/line)

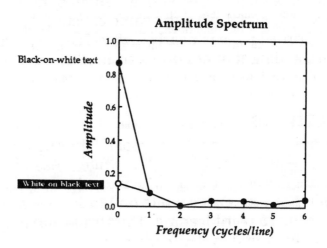

Amplitude Spectrum

Black-on-white text

White on black text

Frequency (cycles/line)

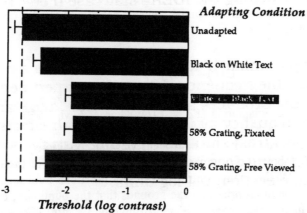

Contrast Threshold at the Line Frequency of the Text

Adapting Condition

Unadapted

Black on White Text

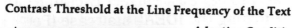 White on Black Text

58% Grating, Fixated

58% Grating, Free Viewed

Threshold (log contrast)

An Image Quality Metric for Digital Letterforms

Joyce E. Farrell and Andrew E. Fitzhugh
Hewlett-Packard Laboratories
P.O. Box 10490
Palo Alto, CA 94303

Applied Problem. Alphanumeric characters often appear distorted when displayed on cathode ray tube devices. These image distortions, referred to as "jaggies", result from undersampling the original high-resolution versions of the characters. To eliminate the aliasing errors that result from undersampling an image, one can filter out frequencies greater than the Nyquist limit - however practical limitation of the display device (gaussian pixel profile and limited intensity range) make the filtering less than ideal. (Kajiya, 1981) Additional image distortions occur because of a mismatch between the filters used to sample a high-resolution image (referred to as convolution kernels) and the pixel point-spread function of the monitor used to display the sampled and filtered image (referred to as the reconstruction kernel).

In practice, the reconstruction kernel of the display is ignored when designing algorithms for sampling and filtering characters. Often this is because the display gamma and pixel point spread function are unknown at the time of character generation. Further, convolution kernels are selected arbitrarily or solely on the basis of minimizing computational time. One would like to select convolution kernels that both minimize computational time and maximize image quality. To solve this problem we need to be able to predict the perceptual effects of various filtering techniques, display reconstruction and viewing conditions.

Metric Approach. One approach to this problem is to develop metrics that can predict how various filtering techniques, display reconstruction and viewing conditions affect image quality. For example, discriminability metrics that predict our ability to detect whether two images are different can be used to determine whether different convolution filters, display architectures or viewing conditions result in perceived differences in character appearance. Such metrics can also be used to select sampling and filtering algorithms that are "optimal" for a given device, where "optimal" is defined as minimizing the perceptual difference between the original high-resolution bit-map character and the sampled and filtered version of that character (Kajiya, 1981)

The experiments described in this paper were designed to evaluate a discriminability metric based on the vector length of the filtered difference between any two images (cf. Mannos & Sakrison ,1974; Kajiya, 1981; Ohtsuka et al, 1988). To calculate this metric, we first filter the difference between two images (eg. an original and it's reproduction) with filters derived from published human contrast sensitivity functions. We assume that human spatial contrast sensitivity can be described by a single contrast sensitivity function of the form

$$C(f) = K(bf)^a * e^{-bf} \qquad [1]$$

where a and b are parameters that depend on the ambient light, pupil size or any other visual factors that may change the state of visual adaptation (Kelly, 1974; Klein & Levi, 1985).

To approximate human contrast sensitivity for equivalent viewing conditions, we derived parameter values by fitting Equation [1] (above) to Campbell's (1968) measurements of the contrast sensitivity function for a 3.8 mm pupil. Of the many contrast sensitivity functions published in the literature, we considered this function to be appropriate on the grounds that the pupil size, estimated from the the DC stimulus luminance in our experiments, is approximately 3.3 mm. (Crawford, 1936). However, the spatial frequency content of the stimuli are such that small changes in the parameters of the contrast sensitivity function do not affect the predictions of the filtered contrast energy difference metric.

The human contrast sensitivity function describes how sensitive a typical or "standard" observer is to the spatial frequency components in any spatial image. Filtering an image with the human contrast sensitivity function then effectively weights the image by how sensitive the visual system is to the spatial frequency components of the image. To quantify the difference between two visually filtered or weighted images we compute the vector length of the filtered difference between the two images. The vector length is based on the Euclidean distance between two points in an n-dimensional space

$$\sum_{i=1}^{n} \sqrt{(A_i - B_i)^2} \qquad [2]$$

where A_i and B_i are corresponding points in images A and B, respectively.

For computational reasons, we use the square of the Euclidean distance or, the squared vector length of the difference. This latter measure is often referred to as the *contrast energy* (Graham & Nachmias, 1978; Nielsen & Wandell, 1988; Watson, Barlow & Robson, 1983). Since the discriminability metric we consider is based on the contrast energy in the difference between any two images, we refer to the metric as the *contrast energy difference* (CED).

Evaluation of the Metric. To evaluate the CED metric, we generated different filtered and sampled versions of the same character by convolving a bi-level master character with different filters and varying the intensity quantization levels of character pixel intensities. Pairs of grayscale characters were simultaneously presented on a monochrome display with known gamma and pixel point-spread functions. Subjects viewed the display from a distance of 12 inches and indicated whether the characters were the same or different in a forced-choice procedure with feedback.

In the first experiment, we investigated the effects of different convolution filters and graylevel quantization. In one condition, referred to as *filters*, grayscale characters were generated using one of four different types of filters (based on box, gaussian, bilinear and sinusoid convolution kernels). In the stimulus condition referred to as *gaussian blur*, grayscale characters were generated using gaussian filters with different widths. And in the stimulus condition referred to as *grayscale quantization*, grayscale characters were generated by filtering each character with a gaussian filter and quantizing the resulting character pixel intensities to 2, 4, 8, 16, 32, 64, 128 or 256 levels of gray.

In our analysis of discrimination performance, we assume performance on the discrimination task depends on the subject's ability to detect the difference between two characters. Discrimination is treated as a special case of detection in which the subject's task is to detect a difference signal. The bigger the difference signal relative to the background noise the more likely it will be detected. Assuming that the distribution of noise and signal have different means but equal variance, we can calculate the d' measure of sensitivity (based on the number of correct discriminations (hits) and the number of false alarms for each character pair).

Figure 1 shows d' plotted as a function of the log CED metric for each character pair presented in each condition of character distortion. For each subject, performance on discrimination tasks (expressed in terms of d') is monotonically related to the CED metric.

In a second experiment, we explored the effects of graylevel quantization on different letters. Again, performance on discrimination tasks was monotonically related to the log CED metric, regardless of the identity of the character.

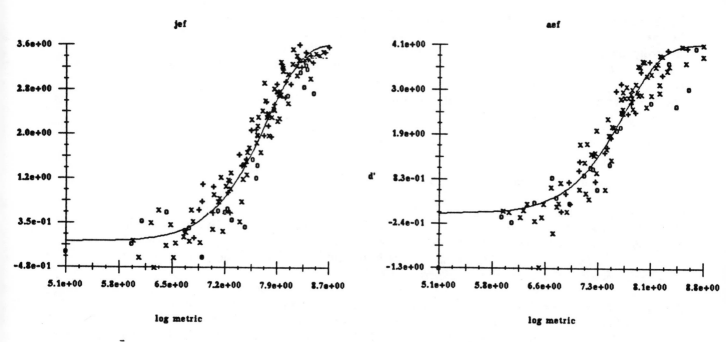

Figure 1. Discrimination performance, expressed as d', plotted as a function of the log *CED* metric for subjects *jef* and *aef*. Data points are parameterized by stimulus condition (x for *filters*, + for *gaussian blur*, and o for *grayscale quantization*). The solid line represents the best-fitting Weibull psychometric function.

Finally, in a third experiment, subjects viewed the stimuli from a distance of 36 inches. To compare metric values for characters presented at different viewing distances, we assume radiant point sources and appropriately correct image intensity by dividing by the square of the viewing distance. Metric values for both 12 and 36 inch viewing distances were calculated assuming a non-linear tranformation of image intensity, I, of the form $R = I^p$. In other words, we make the additional assumption that the visual system has a compressive non-linear response, R, to luminance intensity, I. The exponent, p, that accounts for the data in both viewing conditions is 0.6.

Figure 2 shows d' plotted as a function of the log *CED* metric for character pairs presented at 12 and 36 inches. The fact that the data fall close to one another indicates that discrimination performance at different viewing distances can be predicted by the metric, assuming an initial non-linear transduction of stimulus intensity.

Conclusions. The results of our initial investigations suggest that for some types of stimulus differences, such as contrast quantization errors and gaussian blurring, performance on discrimination tasks is monotonically related to the contrast energy in the filtered difference between two simultaneously viewed images. Moreover, the contrast energy difference metric predicts discrimination performance of different characters viewed at different viewing distances. This result is consistent with the hypothesis that there exists a single psychometric function that can predict the discriminability of different digitized versions of characters when displayed on various devices.

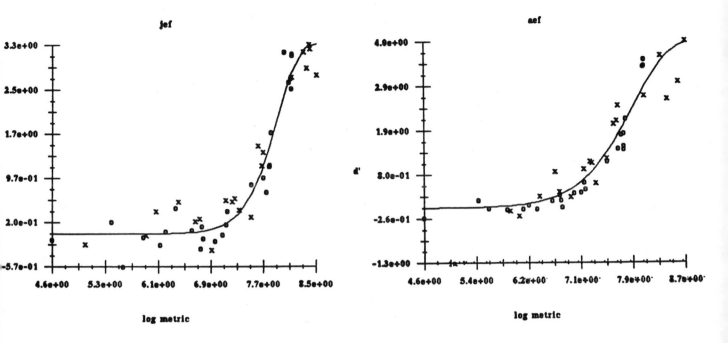

Figure 2. Discrimination performance, expressed as d', plotted as a function of the log contrast energy metric for subjects *jef* and *aef*. Data points are parameterized by viewing distance (x for a viewing distance of 12 inches and o for a viewing distance of 36 inches.) The solid line represents the best-fitting Weibull psychometric function.

References

Campbell, F. W. (1968) "The Human eye as an optical filter" **Proceedings of the IEEE**, Vol. 56, No. 6, pp.1009-1014

Crawford, B. H. (1936) "The dependence of pupil size upon external light under static and variable conditions", **Proceedings of the Royal Society**, (London), Vol. B121, p. 373.

Graham, N., Robson, J. G. & Nachmias, J. (1978) "Grating summation of fovea and periphery", **Vision Research**, Vol. 18, pp. 815-825

Kajiya, J. & Ullner, M. (1981) "Filtering high quailty text for display on raster scan devices", **Computer Graphics**, Vol. 15, No. 3, pp. 7 - 15.

Kelly, D. H. (1974) "Spatial frequency selectivity in the retina", **Vision Research**, Vol. 15, pp. 665-672.

Klein, S. A. & Levi, D. M. (1985) "Hyperacuity thresholds of 1 sec: theoretical predictions and empirical validation", **Journal of the Optical Society of America**, Vol 2, No. 7, pp. 1170-1190.

Mannos, J. L. & Sakrison, D. J. (1974) "The effects of a visual fidelity criterion on the encoding of images", **IEEE Transactions on Information Theory**, Vol. IT-20, No. 4, pp. 525 -536

Ohtsuka, S., Inoue, M. & Watanabe, K. (1988), "Quality evaluation of pictures with multiple impairments based on visually weighted error", **SID Digest**

Nielsen, K. R. K & Wandell, B. A. (1988) "Discrete analysis of spatial-sensitivity models", **Journal of the Optical Society of America**, Vol. 5, No. 5.

Watson, A. B. Barlow, H. B. & Robson, J. G., (1983) "What does the eye see best". **Nature**, Vol. 302, pp.419-422.

THURSDAY, JULY 13, 1989

3:45 PM–5:30 PM

ThC1–ThC5

HIGH DEFINITION AND EXTENDED DEFINITION TELEVISION

William E. Glenn, Florida Atlantic University, *Presider*

PERCEPTION OF "JUTTER" IN TEMPORALLY SAMPLED IMAGES

William E. Glenn, Karen G. Glenn

Department of Electrical Engineering, Florida Atlantic University
Boca Ratan, FL 33431

Abstract

Measurements of contrast sensitivity for "jutter" perception are described as a function of sampling frequency, spatial frequency and velocity for luminance and isoluminant chromaticity gratings.

Motion-picture film and television cameras both sample a motion sequence in time, and display it as a series of fixed images. In film all spatial frequencies in the image are sampled at the same frame rate. It has been found that above about 60 fps, little improvement is obtained by increasing the sampling frequency. In television, because of interlacing, the low spatial frequencies are sampled at 60 fps and the top octave of spatial frequencies in the vertical direction are sampled at 30 fps. Many of the latest bandwidth reduction techniques for high defintion television transmission (VISTA, MUSE, ACTV II, etc.) update higher spatial frequencies at an even lower rate.

Scan conversion between 24 fps film and 30 fps television results in a 12-cycle nonuniformity in motion. Scan conversion from 50 to 60 Hz and vice-versa results in a 10-cycle nonuniformity. It is important to understand the perception of these motion nonuniformities in order to provide some insight into system design.

We have conducted a series of experiments to study the effects of spatial and temporal variables on the perceptibility of motion jutter. Subjectively, motion jutter appears as a nonuniform or "jerky" motion produced by temporal sampling when objects are moving at velocities slow enough to permit visual tracking.

In the first experiment the observer judged which of two successively presented luminance gratings showed motion jutter. As the sine-wave grating moves down the screen it either moves uniformly, or it jumps over varying distances. We observed that the faster the grating's motion, the higher the temporal jutter rate needs to be to be perceptible.

To further study jutter perceptibility, contrast sensitivity measurements were made for luminance gratings and for isoluminance chomaticity gratings. In this way we could measure contrast thresholds for perception of jutter as a function of spatial frequency, sampling temporal frequency and angular velocity.

To study perceptibility of jutter, we used the method of adjustment with the observer adjusting the contrast of a horizontal sine wave grating as it moved vertically across the screen. Thresholds were obtained under various conditions. Three experienced psychophysical observers underwent a series of measurements for luminance gratings, red-cyan (R-Y) and blue-yellow (B-Y) isoluminance chromaticity gratings. In the latter conditions, the adjustments varied saturation of the colors rather than black-white contrast. Prior to the experiment, chromaticity gratings were adjusted by flicker photometry for each observer to minimize the perceived brightness difference between the color pairs.

We observed that viewers were actually more sensitive to the perception of "jutter" at higher velocities. At first this seemed a bit surprising since one would expect more difficulty in tracking higher velocities. We speculated, however, that the observers simply visually tracked the average velocity smoothly, and what he saw was actually the counterphase flicker component of the oscillating grating. At higher velocities, this phase angle was larger and hence easier to perceive.

To explore this hypothesis we corrected the contrast sensitivity measurements by a factor of $\sin \theta/4$, where θ is the phase angle of the jerk. This factor corrects the measurement for the sinusoidal component of counterphase flicker grating. We compared these corrected curves with the MTF's for a stationary grating flickering in counterphase alternation, measured with the same equipment on the same observers. For gratings between about one cycle/degree and 5 cycle/degree the agreement was quite good. At .3 cycles/degree the viewer was more sensitive than predicted and at 10 cycles/degree much less sensitive than predicted by this model. The reason for this discrepancy is as yet unknown. However, one can generalize that a viewer is much more sensitive to jerky motion at low than high spatial frequencies.

From the basic uncorrected data it is possible to derive a spatio-temporal trade-off relationship for jutter perception. They show that the faster the grating's motion, the higher the temporal jutter rate needs to be. However, this analysis assumed that the contrast of the grating did not depend on velocity, and that the image source had 100% modulation transfer function (MTF) at all spatial frequencies.

From our tests it appears that in order to have a sampled image appear to move smoothly at 100% contrast and at all spatial frequencies and velocities for luminance, the sampling frequency needs to be about 1/60 second or higher for the displayed information. It also seems that isoluminance color information can be displayed with minimum sample rates of 1/30 second.

While these rates are required for the display of the information, it does not necessarily mean that all the information transmitted needs to be updated at this frame rate.

In an earlier study (Glenn and Glenn, *Displays, 6:* 202, 1985) we explored the consequences of the loss of limiting resolution for moving objects. It has been known for years that moving objects lose resolution in real cameras due to the integration on film or a television camera tube face. This effect results in a temporal pre-filtering of the displayed information. Its contrast follows the form of a sin X function where X is a function of velocity and spatial frequency. In our earlier study, we compared the apparent sharpness of an image with a 1/60 second integration time with that of images with another form of temporal prefiltering. We found that a temporal prefilter that increased the temporal response of low spatial frequencies at the expense of high spatial frequencies (worse dynamic limiting resolution) actually looked sharper than the prefiltering caused by the usual 1/60 second integration.

This process requires a temporal filter that gives good temporal response below 5 cycles per degree. The contrast of higher spatial frequencies may degrade at higher velocity. The temporal prefilter must keep the contrast below these values at all velocities.

A second way of achieving somewhat similar results is to use motion-adaption processing in the received signal. Many workers use this technique. The resolution is simply "switched" to a lower value when something moves. This technique is less satisfactory than temporal prefiltering since it sacrifices resolution in all directions during motion. Temporal prefiltering only loses resolution in the direction of motion but preserves resolution perpendicular to motion.

The reasonably good confirmation of our hypothesis that "jutter" perception can be explained by counterphase flicker suggests another method of avoiding jutter in moving objects. Processing such as temporal interpolation and other temporal postfilters can remove the low temporal counterphase flicker components that result if detail information is transmitted at low frame rates. They can also remove low temporal frequency, such as beat frequencies when scan converting between frame rates. These routines still require displaying at a high frame rate (60 fps) but can use lower frame rates for transmission.

Interpolation routines have a slight loss in dynamic resolution for moving objects. That is, however, not aa great a loss as with temporal prefiltering. A process with even less loss uses velocity detection and position interpolation of the image of the moving object. This process, however, is probably an overkill. It uses very complex processing to preserve dynamic resolution of very high spatial frequencies. The results of our tests indicate that the difference in dynamic resolution probably does not justify the added complexity.

Theoretically, both temporal interpolation and position interpolation with velocity detection both fail when there is sudden change in velocity. We have not studied the relative visibility of any artifacts that may be introduced under these conditions. These artifacts probably have a low visibility in both cases if the transmitted frame rate for detail is 15 fps or more. They will probably, however, establish a lower limit for the transmitted frame rate of detail information.

113

Visual Perception and the Evolution of Video

C. R. Carlson, J. R. Bergen
David Sarnoff Research Center/SRI
Princeton, NJ 08540

The fundamental objective of advanced television research is to produce beautiful, life-like pictures. The perceptual requirements for producing such vivid images include image size, brightness, contrast, color saturation and purity, and noise and artifact visibility in addition to spatial resolution.

Current Status of the Development of HDTV

Taiji Nishizawa
Science and Technical Research Laboratories
Japan Broadcasting Corporation (NHK)
Tokyo Japan

1. Introduction

NHK has been engaged in the development of HDTV for more than 20 years. Various kinds of equipment necessary for broadcast service have been developed and their characteristics have been improved year by year.

In Japan daily one hour experimental HDTV satellite broadcasting is scheduled to start via BS-2 in June 1989 and practical broadcasting is expected to begin via BS-3 to be launched in 1990.

This paper describes the current status of the development of HDTV technology for broadcasting and other industrial applications.

2. Program production equipment

2.1 Camera

An HDTV hand-held camera using 2/3-inch HARP (High-gain avalanche rushing amorphous photoconductor) tubes has been developed. The sensitivity of it is ten times greater than that of conventional HDTV cameras, and it can be operated at F:2.8 under 200 Lux. This camera provides high quality pictures even in cases with low levels of illumination such as outdoor sporting events, show stages in arenas or theaters.

Solid-state image pickup devices which meet 1125/60 studio standard have also been developed. They include a one-inch format image sensor with two million picture elements, and a laminated CCD image sensor. The next generation solid-state HDTV color camera using one of these has been developed. It is compact and twice as sensitive as conventional models.

2.2 Recording equipment

An open-reel analog HDTV VTR based on a one-inch type C format has commonly been used as the first-generation studio-use HDTV VTR. This system records for a maximum of one hour on a one-inch oxide-coated tape loaded in an 11.75-inch reel. The system uses component signals for input/output.

Another analog HDTV VTR using 1/2-inch cassette is being developed and a number of prototype machines have been completed. Major applications are the distribution of HDTV programs for theaters, educational purposes, medical uses, and printing industries.

A   digital   HDTV   VTR   with   a   recording   capability   of   1.188

Gbits/second has been developed. The recording medium is metal particle tape. The transport mechanism is the one-inch C-type. The sampling frequency of the input/output signals is 74.25 MHz for luminance signal and 37.125 MHz for two color difference signals. This VTR facilitates slow-motion replay.

2.3 Telecine

A frame-rate converter from 24 frames to 60 fields for HDTV telecine has been developed using a motion compensation technology that was studied and used in the development of the HDTV to PAL standard converter. It has been found that the problem of motion judder due to field repetition or 2-3 pull-down in conventional telecine can be solved by using this type of converter so that a natural, smooth motion in the picture can be reproduced on video picture.

2.4 Other equipment

Other equipment such as an HDTV video-matte system, an HDTV to NTSC standard converter and special effects equipment has already been developed.

3. Transmission system

3.1 MUSE family systems

To transmit HDTV signals through 27MHz satellite channels planned at WARC-BS 77, the band-compression technique called MUSE(Multiple sub-Nyquist sampling encoding) has been developed.

For terrestrial broadcasting of ATV(Advanced television) using VHF and UHF channels, NTSC-Compatible MUSE systems have been developed in line with the FCC's tentative decision for the development of ATV made in September 1988. They are the Narrow-MUSE system, NTSC-Compatible MUSE-6 system and NTSC-Compatible MUSE-9 system.

The MUSE and NTSC-Compatible MUSE systems constitute the MUSE family systems. They all have been designed with a wide 16:9 aspect ratio and use the 1125/60 HDTV studio standard.

3.1.1 The MUSE system

The MUSE system employs a 4-field sub-sampling method and reduces the signal bandwidth from 24.3 MHz to 8.1 MHz. Properties of the human visual system are effectively taken into the design of the spatio-temporal sub-sampling, compressing the amount of information to be transmitted.

The MUSE system has been successfully used for a broad range of experimental transmissions by direct broadcast satellite, communications satellites, terrestrial broadcasting systems and CATV systems over the past several years including the international transmissions between Japan and Australia and between Japan and Korea.

A simple MUSE to 525-line standard converter has been developed. In most cases, the picture quality of the 525-line picture converted from MUSE is better than that of pictures originated in the 525-line system. VLSIs for this converter are now being developed.

3.1.2 The Narrow-MUSE system

Narrow-MUSE is a reduced version of the MUSE system. It has 750 scanning lines, in contrast to 1125 for MUSE, and has a baseband width of 4.86 MHz, instead of 8.1 MHz for MUSE, to fit into a 6 MHz transmission channel. It provides better picture quality, given the bandwidth, than compatible systems since there is no NTSC compatibility constraint. This system is designed to be be used for the simulcasting of ATV and conventional television signals.

3.1.3 The NTSC-Compatible MUSE-6 system

The NTSC-Compatible MUSE-6 system can be transmitted in 6 MHz bandwidth channels. For expanding the aspect ratio from 4:3 to 16:9, the letter-box method is used. The high frequency components of the luminance signal are frequency multiplexed into the NTSC signal by using two 2-frame offset sub-carriers to provide additional horizontal details. The vertical high frequency components are transmitted in the masked portions at the top and bottom of the picture. It provides full compatibility with conventional NTSC receivers, but the picture quality obtainable is less than that of the Narrow-MUSE system. For still pictures, a resolution of about twice that of NTSC can be obtained. The maximum vertical resolution is 690 TV lines. 2 channels of digital sound are transmitted.

3.1.4 The NTSC-Compatible MUSE-9 system

The NTSC-Compatible MUSE-9 system is the same as the main 6 MHz channel above, but it uses an additional 3 MHz channel to improve the resolution of the moving portions in the pictures and to provide two additional channels of digital sound.

3.2 Wideband transmission systems

For program material transmission, higher picture quality is required. To meet this requirement, wideband transmission systems such as communications satellites, transportable terrestrial radio links and optical fibers can be used.

In the optical fiber transmission system developed by NHK, the three video components, luminance and two color difference signals, are frequency modulated using different carriers, frequency multiplexed, and then fed to an optical modulator. Using a laser diode, a single mode fiber, and an avalanche photodiode, HDTV signals have been transmitted over 20 km with an unweighted SN ratio of about 50 dB. This system was utilized for the transmission of the Seoul Olympic Games.

3.3 Digital transmission systems

Digital transmission of HDTV is also important not only for long distant transmission but also for distribution to consumers using B-ISDN. So far, the H4 level of ISDN (about 140 Mbits/seconds) is assumed to be the transmission bit rate. DPCM, DCT and vector quantization coding methods are being investigated. Several experimental codecs have been developed and used for field tests and demonstrations.

Development of a digital transmission system for MUSE signals has also been investigated using DPCM coding, aiming at a transmission bit rate less than 100 Mbits/second.

4. Display

Various displays conforming to the 1125/60 studio standard have been developed. Currently, the largest diagonal screen sizes are 41 inches for direct view CRT displays, 400 inches for rear projection displays and 200 inches for front projection displays.

In the near future, an ideal home-use HDTV display may be the rear projection type with a 50-inch screen. Such a display having a screen brightness of 400 cd/m^2 with a simplified automatic convergence adjuster to secure high performance has been developed.

For flat panel displays, a 20-inch display using a gas discharge panel has been developed. A 14-inch liquid crystal color display using amorphous silicon TFTs and a 50-inch projection type liquid crystal display have also been developed. Although these are operated at present with 525-line television, they may be taken as the basis for further development.

5. Receiver

Prototypes of the MUSE decoder using discrete components have already been developed by manufacturers and were demonstrated in June 1988. Development of VLSIs for such necessary components as memories, filters and arithmetic units is progressing on schedule through the joint work of NHK and manufacturers. These VLSIs will be used in MUSE receivers for sale in 1990 when HDTV satellite broadcasting service is expected to start.

A low-noise converter for the satellite receiver using high electron mobility transistors(HEMTs) as the primary stage of the amplifier is widely used and a noise figure of 1.5 dB is now available enabling the use of a smaller antenna. A flat antenna with an array of antenna elements loaded on a printed circuit board is also available on the market. The efficiency of the flat antenna is still being improved.

6. Conclusion

The current status of the development of HDTV technology has been described in this paper. Applications of this technology for both broadcasting and industrial fields are steadily progressing. We believe that HDTV technology can be one of the most innovative in the future information society. We continue our efforts to encourage mutual cooperation in the development of HDTV.

On the Application of Spatio-Temporal Contrast-Sensitivity Functions to HDTV

by

R. Schäfer, P. Kauff and U. Gölz

Heinrich-Hertz-Institut für Nachrichtentechnik Berlin GmbH

Einsteinufer 37, D-1000 Berlin 10

Introduction

Spatio-temporal contrast-sensivity functions have been measured by many psycho-pysicists /1,2/ and the published data have been used by TV engineers for various applications. The most attractive property of these functions is the apparent exchangebility between spatial and temporal resolution, which is utilized by many signal compression schemes (e.g. MUSE and HD-MAC). But it is mostly overlooked that this mechanism is only effective under certain viewing conditions and that there is no reduction in the ability to resolve spatial detail, if the eye can follow the movement. The only mechanism which can really be exploited in this situation is motion blur, which is generated by integration and lag of the camera target /3/. Nevertheless we investigated three different topics related to the spatio-temporal behaviour of human vision which will be discussed in the following.

The first application is a sequential 4-channel-camera system which is similar to a proposal by Glenn /4/, the second topic deals with temporal subsampling in pyramid coding and in the third application we used spatio-temporal contrast-sensitivity functions for 3-dimensional noise weighting in HDTV transmission systems.

4-channel-camera system

A progressive production standard for HDTV implies the necessity to build sequentially scanned HDTV cameras which require a signal bandwidth of 60 MHz in the RGB channels. This causes severe technological problems and especially the signal to noise ratio (SNR) is about 10 dB lower than the one of an interlaced camera with 30 MHz bandwidth.

In order to solve this problem we investigated the 4-channel-camera system shown in Fig. 1. It uses a camera with 4 pick-up tubes, 3 tubes for $R_L G_L B_L$ scanned sequentially with 625 lines and a frame rate of 50 Hz, and one tube for the luminance Y_H scanned sequentially with 1250 lines and 25 Hz frame rate. The four signals are combined in a 3D-signal processor, where three $E_R E_G E_B$ signals with 1250 lines, 50 Hz, 1:1 are generated. In this system vertical resolution is exchanged against temporal resolution, which means that full vertical resolution is only given for temporal frequencies less than 12.5 Hz. However this reduced resolution becomes clearly visible during pan or zoom operations, when the eye can easily follow the motion, whereas this reduction is less critical in the case of uncorrelated movements. Especially vertical or diagonal movements with a velocity of 1 pixel per frame are most critical because in this case the minimum of the vertical temporal resolution is reached. In addition temporal alias distortions occur due to phase errors between the Y_H and the $R_L G_L B_L$ channels.

Subjective tests have been carried out in order to compare this system with an interlaced and with a sequential camera. These tests showed clearly the superior performance of the 4-channel camera compared to an interlaced camera but they also showed that a sequential camera is even better due to the above mentioned artefacts of the 4-channel system.

Another disadvantage of this system is the increased complexity of its optical system which results in problems of registration and loss of light.

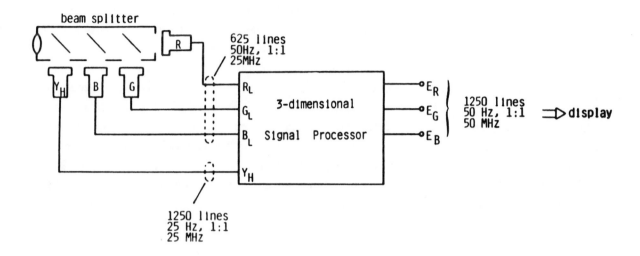

Fig. 1: Blockdiagramm of 4-channel camera system for the generation of sequentially scanned HDTV signals

Temporal subsampling in pyramid coding

In subband coding and pyramid coding spatial frequency channels are generated in which efficient bit rate reduction can be performed. These coding scheemes are often motivated by their similarity to the processing of the visual system itself /5/. Indeed several models of human visual perception have been proposed which contain four to six spatial frequency channels /6/.

Transfering the properties of the spatio-temporal contrast-sensivity functions to this multi channel approach would suggest, that spatial high frequency channels require only a reduced temporal bandwidth, i.e. they can be transmitted at a reduced frame rate. However this is not true due to the above mentioned capability of the eye to follow the motion.

Nevertheless we investigated the possibility to transmit spatial high frequency channels at a lower frame rate in pyramid coding using motion compensated interpolation /7/. In our approach the detail information is only used in those parts of the picture, where the motion vectors are accurately estimated, in all the other parts only the low pass component is used. The psycho-physical justification for this procedure is as follows: In parts of the pictures with correlated motion, which the eye can easily follow, motion estimation is mostly accurate and therefore the subsampled high frequency channels can be reconstructed, whereas in parts of uncorrelated motion, where motion estimation often fails, the eye cannot follow the movements and therefore detail information is not required. Using this approach subsampling factors up to 3 could be used in the high pass channel of a two level pyramid without visible distortions.

Noise weighting

Conventional noise weighting has the crucial disadvantage to work only 1-dimensionally, while video noise is perceived as spatio-temporal statistical fluctuations of luminance and chrominance magnitudes on the display. This disadvantage becomes apparent when noise weighting is applied to HDTV transmission systems like MAC and MUSE, where sophisticated 3-dimensional signal processing is performed at the receiver. Therefore we propose a 3-dimensional noise model, which takes into account the processing at the receiver and the properties of the human visual system. This weighting is performed in a five step operation as shown in Fig. 2. The heart of these operations is the 3-dimensional noise weighting of which only the luminance part has been implemented up to know. For this purpose we approximated Robson's /1/ contrast-sensitivity functions by three horizontal, vertical and temporal Gaussian filters. Although this approximation does not take into account the bandpass characterics of the visual system, it is sufficient for noise weighting, because only the noise power is considered, i.e. the value which results from integration over the entire noise spectrum /8/. With this approach we obtained excellent agreements between theoretical calculations and results of subjective tests for different kinds of spatially, temporally and spatio-temporally filtered noise. In this context the different branches of the HD-MAC system (i.e. 20 msec, 40 msec, 80 msec) have been investigated and the result is that the weighted SNR of HD-MAC is more than 10 dB lower than the one of conventional MAC.

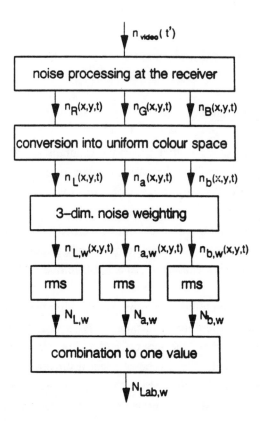

Fig. 2: Concept of 3-dimensional noise weighting

Conclusions

3 different applications of the contrast-sensitivity functions of the human visual system to TV or HDTV have been investigated. As these functions have been obtained under specific viewing conditions which do not take into account the capability of the eye to follow correlated motion in the picture, their applicability is somewhat limited. However very good results have been obtained for noise weighting, where the problem of motion does not occur. For other applications motion estimation and motion adaptive techniques have to be used in order to consider motion artefacts.

References

/1/ Robson, J.G.: "Spatial and Temporal Contrast-Sensivity Functions of the Visual System", JOSA, Vol. 56, No. 8, August 1966, pp. 1141-1142.

/2/ Kelly, D.H.: "Motion and Vision II", J. Opt. Soc. Am., Vol. 69, No. 10, October 1979, pp. 1340-1349.

/3/ Kauff, P.; Schäfer, R.: "Comparison of Sequential and Interlaced Scan for HDTV with Regard to Spatial and Temporal Resolution", Proceeding of International Conference on Consumer Electronics, 8.-10.6.1988, Chicago, USA.

/4/ Glenn, W.E. et al.: "Experimental Camera and Recording System for Reduced Bandwidth HDTV Studio Production", 63rd. NAB Convention, April 14-17, 1986, Las Vegas, USA.

/5/ Martens, J.B.; Majoor, G.: "Scale-Space Coding and its Perceptual Relevance", IPO Annual Progress Report, 1986, pp. 63-71.

/6/ Swanson, W. et al.: "Contrast Matching Data Predicted from Contrast Increment Threshold", Vision Research, Vol. 24, 1984, pp. 63-75.

/7/ Gölz, U.; Schäfer, R.: "Considerations on the Possibility to Exchange Temporal against Spatial Resolution", Proceedings of Picture Coding Symposium, September 12-14, 1988, Torino, Italy.

/8/ Girod, B.: "A Model of the Human Visual System for Irrelevancy Reduction in the Luminance Signal of TV", Fortschrittsberichte VDI, Vol. 10/84, 1988; (in German).

Open Architecture Television

Andrew Lippman
MIT Media Laboratory
20 Ames Street
Cambridge, MA   02139

Abstract:

During the past few years, potentially misdirected effort has been expended developing the notion of high definition television (HDTV) as a system akin to existing television but with approximately twice the spatial resolution horizontally and vertically, and with a slightly higher aspect ratio. System proponents have suggested "production" or distribution systems where the sole difference between current systems and these new ideas is the number of lines and the pixel rate. These efforts have resulted in international discord that may well result in a generation of television systems that impede international interchange of programs, fail to deliver higher quality to the consumer, and retard the technological convergence of computer workstation technology with that of consumer video. In this paper, we suggest that HDTV is an issue of *system architecture* rather than line count, and we explore signal representations that allow for multiple, simultaneous display of disparate TV standards on the same screen. The notions presented here are not fully developed but are fruitful areas of research and study. We suggest that a new generation of television systems be predicated upon sequential component representation of the video sequence rather than a series of frames and that consumer receivers and workstations be optimized for processing this video format.

1. HDTV To Date

The design of a workstation that can display High Definition Television (HDTV) or an HDTV receiver is complicated by the fact that there is no universally accepted definition of the term. In general, the international television industry accepts the notion that HDTV means a signal that comprises approximately twice spatial resolution of current television both horizontally and vertically, and contains a slightly higher aspect ratio. Much of this is based upon work initiated by NHK and reported by Fujio, et al[1] in which a series of subjective experiments seemingly confirmed that maximal viewer acceptance was achieved when the screen approximated the cinematic aspect ratio of 5:3 and the line count approached 1100 lines per picture height. Subsequent to this research, the development of a picture standard with 1125 lines, an aspect ratio of 5:3, and approximately 30MHz of baseband bandwidth was developed and proposed for standardization for television production. Compressed versions of this signal requiring 8.4MHz and eventually, 8.1MHz, call MUSE [2] were proposed as transmission systems for this high bandwidth video system. MUSE has since been developed into a "family" of distribution systems only peripherally related in their basic technological approach to video compression and versions that augment NSTC transmission within the 6MHz band[1] or substitute a new 6MHZ signal[2] have been proposed.

In response to this challenge, Europe launched a program[3] to develop an alternative suited to the 50Hz European environment and consonant with existing European broadcasts. This system is tied to a production standard operating at a 50Hz frame rate and with exactly twice the number of lines as PAL and SECAM broadcasts. The result of this is the potential for a new generation of television systems each parochial in their application in which the immediate future of television fails to achieve universal high quality program interchange or production, and the regional transmission standard divides rather than unifies.

It appears that international discord will prevent any of these systems from being universally accepted, and fundamental problems of conversion of signals that originate at one frame (or field) rate to another will complicate matters rather than ameliorate the situation. Further, the fact that each system initially will employ interlaced scanning means that standards conversion is a complex process requiring high-risk adaptive processing.

The situation is complicated by the fact that there is no basis for assuming that a simple doubling of spatial resolution or widening of the image area will result in widespread viewer approbation. Transmission degradations appear to be more significant than line count, and extra width in the display can appear as lack of height as well as an increment in width.

In terms of computer workstations, HDTV is almost a peripheral issue. Many already operate at line rates equivalent to HDTV, and an increasing number are capable of broadcast television display in a window environment. However, the mere fact of a change in the shape of the CRT envelop can impact future workstation designs, and the ready decoding and incorporation of HDTV signals into the workstation environment is an important criterion for the future. Indeed, much of the processing power of a workstation may well be required for decoding an HDTV distribution standard, and the distinction between workstations and television receivers will diminish. The internal circuitry may be similar but the operational facade presented to the owner may depend on whether the system is a desktop engineering station or an entertainment viewing device.

2. The Open Architecture Receiver

To unify the approach to HDTV and to exploit the commonality between consumer receivers and professional workstations we propose an Open Architecture Television Receiver (OAR.) The OAR consists of a set of image decoders each designed to accept a local distribution standard and translate it into a common, Digital Intermediate Format for transport along an internal image bus. This bus can contain video information as well as act as communications system for components of a normal workstation or personal computer.

A variety of display systems may be attached to the bus, each of which will translate the intermediate format to a sequence of raster scan images suitable for the particular display technology associated with the adapter itself. The line and frame rates need not be simply related to either the video transmission rate or the internal refresh rate of the display, and the entire system is optimized for the simultaneous display of multiple, inconsistent video standards. To a great extent, the system goals can be described quite simply by a diagram showing a prospective display,

---

1. These are called EDTV systems, for Extended Definition Television. In EDTV systems, the normal receiver can decode a normal program, but a special receiver can decode augmentation information that results in a higher definition image.

2. New use of an existing channel allocation is usually referred to as a "simulcast system" because it is presumed that a normal NTSC version of the program would be distributed through existing channels and new ones would be allocated for this new signal. The new signal occupies precisly one television channel.

3. Eureka-95, dedicated to the development of a uniquely European HDTV system.

Figure 1, where two windows, each originating at a different frame rate are overlaid on a display operating at a third refresh rate that is not either of the incoming video standards.

A candidate OAR architecture is shown in Figure 2. In the figure, a set of adapters accepts video from sources as diverse as over-the-air tuners, compressed video decoders, and personal computer graphics generators. These are each translated into the internal format for presentation to the video bus of the system. Attached to this bus is a set of displays each containing an interpolator to translate the bus format into that required for the line and frame rate of the particular display unit to be used. Of special importance is the notion that multiple displays may be used in the same system and each may have different resolutions and frame rates.

The system architecture can be adapted to scalable bus architectures that have been proposed for multi-processor interconnects and networks.

3. Digital Intermediate Format

Of key importance in this type of system is the definition of the internal digital video standard. We suggest a digital intermediate format (DIF) that is of a subband transformation of the time series of video frames. This format results from a spectral decomposition of the video frame sequence into a set of spatio-temporal components by filtering and subsampling.[3] In particular, the image sequence is filtered by a tree-structured array of separable bandsplitting filters alternately applied in horizontal, vertical and temporal orientations, and the resulting subbands are then subsampled to reflect the new information content of each band.[4] [5] [6] A suitable decomposition is shown in Figure 3. Other alternative representations such as transforms, motion fields and keyframes, or even explicit picture element representations are candidates, but the spectral components offer several advantages, as listed here:

☛ **Direct Interface to High Bandwidth Production Standards:** The DIF can serve as a digital representation of the video signal for high bandwidth production systems. It can interface directly to video recording equipment and optical networks.

☛ **Multi-Standard:** DIF allows images derived from diverse standards to be simultaneously displayed.

☛ **Multiple Displays:** Various superficially incompatible displays can be attached to the bus, as described above.

☛ **Scalability:** The DIF can be upgraded to suit a variety of system costs and architectures.

☛ **Data Compression:** The component representation facilitates representation of each component at different signal-to-noise ratios, thus allowing inherent compression that is not available in an explicit frame series video representation.

☛ **Graceful Degradation:** System overload can be handled by deleting components rather than by deleting frames. This can happen in cases where multiple translucent windows are displayed.

More recently, Schreiber[7] has proposed that a subband decomposition of the video signal based upon a temporal rate of multiples of 12Hz can serve as a unifying standard for international television program exchange. In this "Friendly Family," diverse standards such as 24 fps film, 50Hz and 60Hz video are all transformed to the subband representation where the particular origination standard determines what bands have information content and which are empty. The subband representation provides a format whereby these seemingly incompatible standards may be processed in common and presented for display. The DIF extends this notion to include computer generated video and to accept displays operating at none of the incoming standards.

4. Conclusion

In this paper, we present an architecture to television systems and workstations based on a common bus and a digital intermediate format for the video signal. This bus allows ready extensibility, and the signal representation facilitates intermixing diverse frame and line rates. We suggest that this is an economic solution to the next generation of television systems that avoids *a priori* definition of a single production, transmission or display standard and replaces it with an extensible system architecture. Economics and applications can dictate the particulars of any implementation.

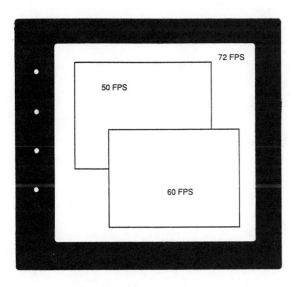

Fig. 1 Multiple Inconsistent Television Standard

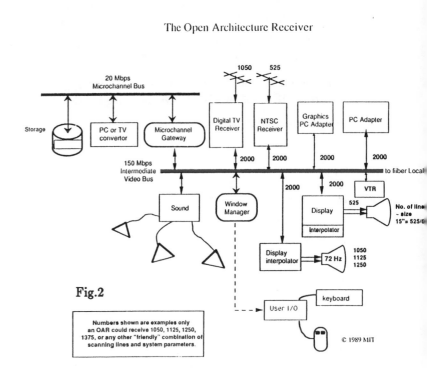

The Open Architecture Receiver

Fig.2

Numbers shown are examples only
an OAR could receive 1050, 1125, 1250,
1375, or any other "friendly" combination of
scanning lines and system parameters.

© 1989 MIT

Pyramid Decomposition of a Frame Pair

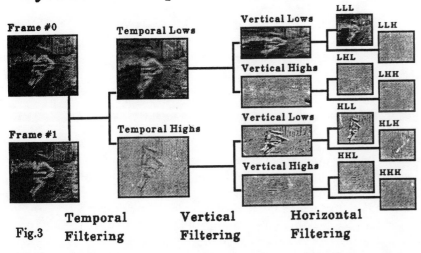

Fig.3 Temporal Filtering   Vertical Filtering   Horizontal Filtering

REFERENCES

1. Fujio, T., et al, High Definition Television, NHK Monograph, Number 32, June 1982.

2. A Single Channel HDTV Broadcast System -- the MUSE, Ninomiya., y., et al, NHK Lab note, Sept 1984.

3. Lippman, A., and Butera, W., Coding Image Sequences for Interactive Retrieval, Proc ACM, Forthcoming.

4. Vaidyanthan, P. P., Quadrature Mirror Filter Banks, M-Band Extensions, and Perfect Reconstruction Techniques, ASSP Magazine, July 1987.

5. Karlsson, G., and Barnwell, T., A Procedure for Designing Exact Reconstruction Filters for Tree Structured Subband Coders, IEEE conf on ASSP.

6. Karlsson, G., and Vetterli, M., Subband Coding of Video Signals for Packet Switched Networks, Proc SPIE Conf on Visual Communications, Oct, 1987.

7. Schreiber, W. F., Friendly Family of Television Standards, NAB, 1988.

FRIDAY, JULY 14, 1989

9:00 AM–11:45 AM

FA1–FA7

COLOR RENDERING

William B. Cowan, University of Waterloo, *Presider*

Device Independent Color Reproduction

Maureen C. Stone
Xerox Palo Alto Research Center
Palo Alto, CA 94304

Abstract: Computer systems can be used to interconnect many different types of color devices: monitors, printers, film and video recorders. Techniques based on CIE standards provide a degree of "device independence" to color reproduction in these systems. However, reproducing tristimulus values, pixel by pixel, will not result in acceptable color reproduction for images or related color sets. Differences in gamut and appearance characteristics make an additional transformation step essential, a step we call *gamut mapping*. While this model has proven useful for color reproduction across media [11], there are still research problems left to be solved before it can be fully realized. This paper will discuss these issues and also some of the inherent limitations of this approach to color reproduction.

1 Introduction

Computer systems can be used to interconnect different color producing devices such as monitors, printers, film and video recorders. Different media are often available on the same computer system or network. Each different medium has it's own color representation, typically, RGB (Red, Green and Blue) for additive devices such as monitors, and CMYK (Cyan, Magenta, Yellow and Black) for subtractive devices such as printers. Converting between device-level color representations is difficult, and the number of converters needed approaches the square of the number of devices connected to the system.

With tools for colorimetry becoming faster and cheaper, and with digital systems providing direct control of the color of each pixel, device colors can be expressed in some standard form, typically CIE tristimulus values or some color system derived from tristimulus values such as CIELAB or CIELUV. representation. A procedure, called a *characterization* can be created for converting between tristimulus values and device color coordinates for every medium on the system. The characterization provides a quantitative way to reproduce any tristimulus value within the capabilities of the device. It also provides a common way to define the set of all possible colors each device can produce, its *gamut*.

Given the characterizations, any color generated on one medium can now be reproduced on another by reproducing its tristimulus values, assuming the color lies inside the target device's gamut. However, reproducing tristimulus values pixel by pixel does not produce acceptable reproductions of related color sets or images. An additional transformation of the tristimulus values is needed to compensated for differences in viewing conditions and gamut limitations, a process we call *gamut mapping*.

There are unsolved problems associated with both device characterization and gamut mapping. Characterization problems focus on the speed and accuracy of the conversion between tristimulus values and device coordinates, as discussed in Section 2. Creating effective gamut maps currently requires human judgement and iteration. While some principles controlling these maps can be described (Section 3) human judgement will always be required at some level. The paper concludes with a discussion of the advantages and disadvantages of this model for color reproduction.

2 Device Characterization

A characterization forms a conversion between device coordinates and and some external standard such as CIE tristimulus values. That is, it creates a mapping between an n-dimensional device coordinate system to an m-dimensional standard representation. In the case of printers, n is typically 3 or 4, though higher dimensions are possible if custom ink sets are used. For the CIE standards, $m = 3$. The inverse of this function must also be provided to map from tristimulus values to device coordinates for rendering on the target medium.

There are two approaches for generating this function: modeling and sampling. Color monitors can be modeled with a good degree of accuracy and efficiency [1]. Most subtractive systems, however, are best characterized by sampling and interpolation because the models are complex and the parameters for the models are difficult to obtain. Furthermore, new technologies are being developed very rapidly and sampling is technology independent.

While it is theoretically possible to characterize a device completely by sampling, first performing a level of device-specific calibration can improve the accuracy of the function and reduce the required sampling density. Once a representative set of samples has been produced, they are measured to form a piecewise function of order n for mapping from device coordinates to tristimulus values. There will be m such functions, one for each of the standard parameters. If the sampling is sufficiently dense, these can be piecewise linear functions. However, as measuring large numbers of samples is tedious without special hardware, higher order functions combined with fewer samples are often used commercially [7][5].

The mapping from device coordinates to tristimulus values and back must be performed for every different color in the scene (a cache should be used to avoid performing it for every pixel) so an efficient implementation is crucial for good performance. Hardware implementation of sampling and interpolation is well within the state-of-the-art for current technology [8], making it potentially an optimal technique for all devices.

Once a characterization is established, regular evaluation is necessary to be sure that it remains accurate. The required stability for practical systems has yet to be fully defined, but experience indicates that many low cost devices will need frequent adjustment. This observation suggests that highly automated characterization methods which are efficient enough to perform regularly will be required for practical systems.

3 Gamut Mapping

Gamut mapping transforms the tristimulus values * that describe the image to tristimulus values suitable for rendering the image on the target device. This transformation is necessary to accommodate the differences in appearance of different media. It is important to emphasize that gamut mapping in some form is essential; reproducing tristimulus values across media will not produce an acceptable reproduction except in very special circumstances. While initial efforts at gamut mapping have focused on linear transformations of tristimulus values, analysis of effective engineering solutions to cross media rendering strongly suggest that nonlinear transformations will be required as well.

Gamut transformations can either map the image gamut, γ_i onto the target device gamut, γ_d, or they can map one device gamut into another: $\gamma_{d1} \Rightarrow \gamma_{d2}$. This second form assumes that the image was originally designed for γ_{d1}, so $\gamma_i \subseteq \gamma_{d1}$. A transformation that maps $\gamma_i \Rightarrow \gamma_d$ can be constructed to optimize the reproduction of just the colors in γ_i. However, a generic map from γ_{d1} to γ_{d2} can be applied to all images designed for $d1$ without generating out-of-gamut colors. Experience has indicated that such mappings can be constructed to give snap-shot quality reproduction [9]. In constructing gamut mappings, the critical factors have been found to be: the neutral axis, saturated colors and out-of-gamut colors.

The neutral axis contains the achromatic colors shading from black to white **. The tristimulus values corresponding to black and white, called the black point and the white point, respectively, are different for different media. The distance between these extremes, the *contrast*, varies as well. Traditional practice emphasizes that it is important to align the neutral axis of the original with the reproduction. This means that the neutral axis of the image gamut should be transformed to align with

---

    * In this section, tristimulus values are assumed as the standard representation although the discussion applies equally well to derivatives of tristimulus values such as CIELAB and CIELUV

    ** Black and white here are used as appearance terms: black is the darkest achromatic color, and white is the lightest.

the neutral axis of the output gamut and scaled so that the original contrast is maintained. The open question is how to space the gray values along the neutral axis, that is, what is the *tone reproduction curve*.

Traditionally, the $\log_{10}$ of the luminance of the original is mapped to film or print density in the reproduction, where the density, $D = \log 1/R$ with R defined as the reflectance (or transmittance) of the medium. [4, pages 49–62]. However, it is not clear that this is the correct model when the original exists only on a color monitor. Some authors [9] recommend mapping L^*_{monitor} to L^*_{printer}. Experience indicates that the correct mapping is highly image dependent, though the mapping based on L^* can provide good reproduction for a large class of images.

Highly saturated colors (when appropriate) are considered a mark of good reproduction. Monitor colors are much more saturated and brighter than those available on prints or even transparencies. These colors lie far outside the printer or film gamuts, making them impossible to reproduce in a colorimetric sense. It is important, therefore, that the gamut mapping preserve the appearance of saturation as much as possible. This can imply the use of a non-linear transformation to warp the source gamut to fit with the extremes of the destination gamut.

When mapping $\gamma_i \Rightarrow \gamma_d$, it may be globally advantageous to use a mapping that leaves some of the image colors outside of the gamut of the destination device and approximate these colors individually before rendering them. Projecting them to the surface of the destination gamut, either orthogonally or along the line of constant hue or brightness will produce an acceptable result so long as the color is not very far outside of the gamut. When the color lies far outside the destination gamut, experience suggests that a satisfactory approximation cannot be found without higher-level knowledge of scene content.

4 Discussion

Standards based on characterization by tristimulus values have been published by Xerox [2] and are being considered by ISO [6]. While neither have a specific provision for gamut mapping, the Xerox Color Encoding Standard does include a field of arbitrary length for an appearance "hint" which can be used to encode information needed to create a gamut transformation. This field was deliberately left unspecified as the precise form of the information needed is not yet understood.

This model is also being proposed for the commercial film and video industry to provide a standard description for film, video, and computer generated imagery [3]. Some form of color transformation to accommodate cross media appearance are expected to be included in the definition of the standard.

There are inherent limitations with this model. First, tristimulus values are not a very good representation for expressing appearance relationships. Other representations, such as light sources and surface reflectances [12] should be more appropriate. Second, representing image colors as the tristimulus values of individual pixels provides no information about intercolor relationships. For example, a set of colors shading from light to dark may be a color vignette or a shaded object. Such a set of colors should be treated so there is no hue shift along the gradation. However, it is difficult, if not impossible, to accurately derive this relationship from the pixels alone. Ultimately, many important perceptual judgements about the reproduction are defined by scene content and must be represented at a higher level than pixels and tristimulus values.

5 Conclusion

The use of colorimetric characterization for different forms of digital devices is an important step towards device independent color. However, tristimulus values alone provide inadequate information for good color reproduction. More work is needed to determine how to express the missing information.

Gamut mapping is one model for utilizing this information. Different mapping parameters can be used to express important appearance information such as contrast and saturation. Gamut transformations and sampled calibrations can easily be implemented in high performance hardware. While other approaches may provide more accurate in the long run, this approach provides an important first step away from the strictly craft-oriented approach to color reproduction.

This model was originally applied by the author and her colleagues to the problem of reproducing computer graphics imagery for journal publication [10] and further work in this area is planned. In this application, it is interesting to consider that a high level description of the scene already exists in digital form. For 3D graphics, this representation will consist of surfaces and light sources. For 2D graphics, a high-level representation is not guaranteed, but may possibly be derived from the tools used to select the colors. The colors may have been selected together as part of a palette, for example, or color gradations are often specified as an algorithm for blending two colors. Future systems may well be designed to take advantage of this information when considering the problem of cross media rendering.

6 Acknowledgments

Much of this work has been performed in collaboration with William B. Cowan and John C. Beatty of the University of Waterloo, Waterloo, Ontario, Canada. I would also like to thank Richard J. Beach for his support both as my manager at Xerox PARC and as editor-in-chief of ACM SIGGRAPH.

References

[1] W.B. Cowan and N.L. Rowell, "Phosphor Constancy and Gun Independence in Color Video Monitors." *Color Research and Application*, **11**(Supplement) 1986, S34-38.

[2] "Xerox Color Encoding Standard." XNSS 288811, March 1989. Xerox Systems Institute, 475 Oakmead Parkway, Sunnyvale, CA 94086.

[3] L.E. DeMarsh, HDTV Production Colorimetry. Presented informally at the ISCC symposium on *CRT-to-Hard Copy in Color,* May 8-10, 1988.

[4] R.W.G. Hunt, *The Reproduction of Color,* 4th edition, John Wiley & Sons, New York, 1987.

[5] Imageset, San Francisco, CA, private communication.

[6] ISO/IEC JTC 1/SC 19, Text and Office Systems, Addendum on Colour, working draft, July 18. 1988.

[7] B.J. Lindbloom, "Accurate Color Reproduction for Graphics Applications," to appear in *Computer Graphics,* **23**(4), August, 1988.

[8] A.W. Paeth, "Algorithms for Fast Color Correction," to appear *Proceedings of the Society for Information Display,* **30**(2), July 1989.

[9] W.L. Rhodes and M.G. Lamming, "Towards WYSISYG Color," Xerox Palo Alto Research Center, Electronic Document Laboratory, April 1988, EDL-88-2.

[10] M.C. Stone, W.B. Cowan, J.C. Beatty, "A Description of the Reproduction methods used for Color Pictures in this Issue of *Color Research and Application, Color Research and Application,***11** (Supplement), June 1986, S83-S88.

[11] M.C. Stone, W.B. Cowan, J.C. Beatty, "Color Gamut Mapping and the Printing of Digital Color Images," *ACM Transactions on Graphics,* **7**(4), 1988, 249-292.

[12] B.A. Wandell and D.H. Brainard, "Towards Cross-media Color Reproduction," to appear elsewhere in this proceedings.

Towards cross-media color reproduction

Brian A. Wandell and David H. Brainard
Psychology Department
Stanford University
Stanford, CA 94305

1. Introduction

Suppose that we wish to compare the color appearance of an image displayed on monitor and a reproduction of the image on a printed page. We could place the monitor and printed image side-by-side and let an observer judge the colors of each. The difficulty with this arrangement is that the ambient illumination interferes with the color appearance of the monitor image. To see the monitor image clearly, the observer would like to turn down the room lighting. But in this case, the observer will be unable to see the printed image. As the room lights are turned up, the printed image becomes visible but the monitor image becomes washed out.

The conflict between the appropriate ambient lighting conditions for viewing monitor and printed images illustrates one of the primary challenges of cross-media color reproduction. We believe that this conflict is best understood by recognizing that the visual system interprets images as illuminated surfaces and adjusts to the ambient illumination to keep the color appearance of surfaces constant. Thus, for a wide range of ambient illuminations, the color appearance of a printed image does not change much.[1,2,3] Although the light reflected to the eye from a printed page varies with changes in illumination, the visual system adjusts to these lighting changes so that the color appearance of surfaces remains approximately constant. For a monitor image, on the other hand, the effect of the visual system's adjustment is quite different. The light coming from the monitor is (except for the glare reflected from the monitor's screen) independent of the illumination. When the room lights are turned down, the visual system interprets the monitor image as a set of illuminated surfaces. As the room lights are turned up, this interpretation is contaminated and the appearance of the monitor image changes.

The above example illustrates that when we reproduce a monitor image with a printed image, our goal should not be to have the printed image appear identical when the two are viewed side-by-side. Rather, we want the visual system's interpretation of the surfaces implicit in both the printed and monitor images to be the same when each is seen under its preferred illumination. The purpose of this paper is to outline a procedure for trying to arrange this sort of *surface color match*.

Our procedure for achieving surface color matches consists of two parts. The first is to analyze a monitor image to determine what surfaces the visual system is likely to find implicit in it. The second is to approximate any given surface reflectance function on a piece of paper using a color printer. We describe each of these separately in the next two sections. To reproduce a monitor image, we first determine the implicit surfaces from the monitor image and then reproduce these surfaces on the printed page.

2. Monitor image analysis

How do we determine the implicit surfaces of a monitor image? There are two separate cases. The first is when the first case the monitor image is a simulation of the appearance of a scene where the surface reflectance functions and illuminant spectral power distributions are known. This is the case in for images rendered using the Stanford *Color Analysis Package (CAP)* software system,[4,5] where the hardware-dependent display values for monitor images are actually constructed from internal location-by-location representations of surface and illuminant properties. If the underlying surface representation is available, then we assume that the observer's visual system correctly interprets the surface whose appearance was simulated and use the surface representation directly.

It has not escaped our attention that some color processing systems do not use the surface and light representation provided by *CAP*. The second case is when the monitor image is specified in CIE XYZ tri-stimulus coordinates.[*] This is currently the more common case, and it is necessary to convert the hardware-independent representation to a reasonable estimate of the implicit surfaces. Here we describe a method that is based on the use of linear models to represent the implicit surface reflectance functions and illuminant spectral power distributions. Elsewhere, we and a number of our colleagues have described the advantages of using linear models to represent surface reflectance functions and illuminant spectral power distributions.[5,6,7,8,9,10,11] We assume that a known set of tri-stimulus coordinates, typically those corresponding to the monitor white point, are interpreted by the visual system as white paper under the implicit illuminant. Denote these tri-stimulus coordinates by X_w, Y_w, Z_w. Denote the surface reflectance function of white paper by $W(\lambda)$. Furthermore, we assume that the ambient illumination is one of the daylight spectral power distributions defined by the CIE 1971 daylight illuminant linear model. The basis functions of this model, denoted here as $E_i(\lambda)$, are listed in table V(3.3.4) of Wyszecki and Stiles' book.[12] Thus the spectral power distribution of the implicit light in the monitor image is determined by three parameters, ε_i, as $E(\lambda) = \sum_{i=1}^{3} \varepsilon_i E_i(\lambda)$. From the equation

$$X_w = \sum_{i=1}^{3} \varepsilon_i \left[\sum_{\lambda} \bar{x}(\lambda) W(\lambda) E_i(\lambda) \right] \qquad (1)$$

and two parallel linear equations for Y_w and Z_w, we can solve for the three unknown implicit light parameters.

Similarly, if we assume the surface reflectance at each image location is one of the functions from Parkkinen et al.'s basis set,[10] then the implicit surface reflectance function is determined by three parameters, σ_i, as $S(\lambda) = \sum_{i=1}^{3} \sigma_i S_i(\lambda)$. From tri-stimulus values at each image location, X, Y, and Z, and from the equation

$$X = \sum_{i=1}^{3} \sigma_i \left[\sum_{\lambda} \bar{x}(\lambda) E(\lambda) S_i(\lambda) \right] \tag{2}$$

and two parallel linear equations for Y and Z, we can solve for the three unknown implicit surface parameters.

3. Surface reflectance analysis

When we place ink on the printed page, we control the surface reflectance functions of the printed image. To reproduce the surfaces implicit in the monitor image, we generate the hardware control signals to print a surface that approximates the desired reflectance. In this section, we describe a procedure to do this for a particular modern commercial printer, the Hewlett-Packard PaintJet. The PaintJet generates colors using an error-diffusion dithering algorithm that places four different types of inks adjacent to one another on the printed page.

Our procedure uses a linear model to describe the gamut of surfaces that can be produced with the PaintJet. To create this linear model, we began with measurements of a large number of surface reflectance functions produced by the printer.[**] We expressed each measured surface reflectance as the sum of two terms

$$S(\lambda) = W(\lambda) - A(\lambda) , \tag{3}$$

where $W(\lambda)$ is the reflectance function of the white paper and $A(\lambda)$ is the change in surface reflectance caused by the inks. We then computed a linear model to describe the set of measured changes. To do this, we wrote removed the mean change from the data set. If we call the mean change $A_0(\lambda)$, then the linear model describes the observed changes through the equation

$$A(\lambda) \approx A_0(\lambda) + \sum_{i=1}^{i=N} \alpha_i A_i(\lambda). \tag{4}$$

The $A_i(\lambda)$ are the fixed basis functions of the linear model. These were determined from the data using the singular value decomposition. The α_i are parameters that characterize a particular change $A(\lambda)$. The relation between the linear model and the actual surface reflectance function is given by

$$S(\lambda) \approx W(\lambda) - A_0(\lambda) - \sum_{i=1}^{N} \alpha_i A_i(\lambda). \tag{5}$$

As we use more basis functions (i.e. as N increases), we can approximate the data with arbitrary precision. We have found that for the 193 samples from the H-P PaintJet, which were selected to span the printer's range, the surface reflectance functions could be closely approximated using $N = 4$ basis functions. With four basis functions the typical root mean squared error of linear model fit to the true surface reflectance functions was less than two percent. This deviation is about equal to the size of the measurement error.

As we shall show at the talk, for the H-P PaintJet, the weights α_j can be related to the relative proportions of the four inks that are placed on the page and thus to the hardware control signals sent to the printer. To obtain a desired surface reflectance function on the page, we first solve for its best least-squares representation α_i within the linear model (equation (3)). We then convert the linear model representation to the hardware control signal that will produce the closest match that can be generated by the printer.

4. Summary

When matching the color of a monitor image and a printed image, the two images are normally viewed under asymmetric conditions. Under such asymmetric matching conditions, location-by-location tri-stimulus matches do not generate appearance matches.[13] We suggest analyzing the monitor image in terms of implicit surfaces and illuminants. We use the printer to reproduce the implicit surface reflectance functions of the monitor image.

Our remarks also have implications for the format of digitally stored colored images. The methods we have described are best implemented using image processing software that permits representation of images in terms of surfaces and illuminants. The use of linear models to represent these functions offers an efficient method for coding surface reflectance and illuminant data.

Footnotes

ACKNOWLEDGMENT: Supported by NASA (ARC) grant NC2-307, and NEI grant EY03164, both to Stanford University. We thank R. Motta and J. Farrell for their help and support.

* For simplicity of exposition we will assume that the monitor image is specified in terms of CIE XYZ tri-stimulus coordinates. It is straightforward to convert a specification in any reasonable color coordinate system to this standard coordinate system.

** Our colleague at Hewlett-Packard, Ricardo Motta, kindly provided us with measurements of the spectral reflectance functions of 193 different colors produced on a piece of white bond paper.

REFERENCES

1. R. Evans, "Visual Processes and Color Photography," *Journal of the Optical Society of America*, vol. 33, no. 11, pp. 579-614, 1943.

2. D. B. Judd, "Hue saturation and lightness of surface colors with chromatic illumination," *J. Opt. Soc. Am.*, vol. 30, pp. 2-32, 1940.

3. K. McLaren, "Edwin H. Land's contributions to colour science," *JDSC Proceedings*, vol. 102, pp. 378-383, 1986.

4. D.H. Brainard and B.A. Wandell, "The Color Analysis Package," *Color Research and Application*, in press.

5. B. A. Wandell, "The synthesis and analysis of color images," *IEEE PAMI*, vol. PAMI-9, pp. 2-13, 1987.

6. J. Cohen, "Dependency of the spectral reflectance curves of the Munsell color chips," *Psychon. Sci*, vol. 1, pp. 369-370, 1964.

7. D. H. Brainard, B. A. Wandell, and W. B. Cowan, "Black light: How sensors filter spectral variation of the illuminant," *IEEE Trans on Biom. Eng.*, vol. 36, no. 1, pp. 140-149, 1989.

8. G. Buchsbaum, "A Spatial Processor Model for Object Color Perception," *J Franklin Institute*, vol. 310, no. 1, pp. 1-26, 1980.

9. L. T. Maloney, "Evaluation of linear models of surface spectral reflectance with small numbers of parameters," *J. Opt. Soc. Am. A*, vol. 3, no. 10, pp. 1673-1683, 1986.

10. J. P. S. Parkkinen, J. Hallikainen, and T. Jasskelainen, "Charcteristic spectra of Munsell colors," *J. Opt. Soc. Am.*, vol. 6, no. 2, pp. 318-322, 1989.

11. D. B. Judd, D. L. MacAdam, and G. W. Wyszecki, "Spectral distribution of typical daylight as a function of correlated color temperature," *J. Opt. Soc. Am.*, vol. 54, p. 1031, 1964.

12. G. Wyszecki and W. S. Stiles, *Color Science,* John Wiley and Sons, New York, 1982.

13. G. Wyszecki, "Color Appearance," in *Handbook of Perception*, ed. J. P. Thomas, p. Chapter 8, 1986.

COLOR GAMUT MATCHING FOR HARD COPY

JOHN MEYER, BRIAN BARTH
HEWLETT-PACKARD LABORATORIES
1501 PAGE MILL ROAD, PALO ALTO
CALIFORNIA, 94303

ABSTRACT

The input data from a scanner or computer generated image will often contain colors which are outside the printable gamut of a hard copy device. This paper describes a method for obtaining a perceptual appearance match between the original image and the hard copy output under these conditions.

INTRODUCTION

The proliferation of color hard copy devices for use with color data sources such as desktop scanners or computer generated images has put renewed emphasis on an old problem; the inability to print out of gamut colors. This is the nature of any color reproduction technique which relies on a trio of color primaries to span the visible color space. The severity of the problem is directly but not linearly related to the volume enclosed by the primaries and their various combinations in the space used for color measurement. It is generally true that photographic and CRT gamuts exceed those accessible to hardcopy devices although it is not uncommon for the reverse to occur, in which case the printer has the opportunity to enhance the final result. The focus here is on the typical problem of transforming the non-printable colors of an image into printable colors and not on correction or enhancement of poor quality images. In other words, the desired appearance of the image is assumed to be specified correctly by the data.

THE GOAL: APPEARANCE MATCHING

The transformation from non-printable to printable colors often goes by the name of gamut compression in the sense that the representation of the image on a device with a smaller gamut implies less visual separation between the colors. The approaches to the problem fall into two groups: either image colors that are within the gamut are adjusted to "make room" for those that have to be included from outside the printer boundary, or those external colors are mapped, according to some algorithm, to specific points on the printer gamut surface. The first method involves an overall compression whereas the second clips the non-printable colors to the boundary in some way that minimizes the deg- radation of the resultant output. In either case, the question arises as to what should be the objective of such a transformation. Ideally, the original and reproduced images should appear the same. This is to be taken to mean that although there are measureable differences between the original and reproduced colors, the relationship of any color in the image to the overall image will appear to have been preserved. For example, neutrals should be reproduced as neutrals, fully saturated colors should be printed at the limit of the printer's saturation capability, maximum use should be made of the printer's lightness range and hue shifts should be minimal. Any form of enhancement which makes deliberate, interpretative changes of an artistic nature will, of course, obscure the effects of any gamut compression scheme and has been avoided in the approach described below.

EXISTING METHODS

Both photography and graphic arts color separation processes have well developed methods for achieving gamut compression. In each case, considerable use is made of tone scale modification along with varying degrees of color correction for purposes of achieving accurate hues. These techniques have obviously worked very well, but in the case of the graphic arts the intervention of a skilled operator is necessary for even modest performance. The photographic process works best under controlled lighting conditions so that the exposure range of the film is not exceeded and the subject matter is illuminated to make optimum use of the built-in contrast of the film. In each case, the process is interactive and often iterative in nature. The method described in this paper is designed to be very close to automatic in its application.

Recently published work at XEROX PARC [1,2,3] combined colorimetry and digital image processing to control gamut mapping. The overall purpose was the preparation of color separations for offset printing of a selection of computer generated images for a special issue of Color Research and Application [2]. Although the project spanned the entire process of printing color images, special emphasis was placed on the gamut mapping step as the link between the calibrated input and output processes. The processing scheme consisted of three sequential transformations. The first, involving translation, scaling and gray axis rotation constituted the main part of the operation. Its primary function was to adjust the neutral axis of the image so that it was both aligned with and approximately the same size as the gray axis of the destination device. Next came an adjustment fcr highly saturated colors called the umbrella transformation. This desaturated the image by moving the output primaries toward the white point. Finally, the remaining out-of-gamut colors were clipped to the gamut surface thus avoiding overcompression of the image by attempting to locate all colors within the gamut boundary. This three-step technique still requires operators knowledgeable in color science and printing for good results.

Several theses at MIT have addressed the issue of gamut compression [e.g. 4,5]. In his Ph.D. thesis Jason Sara [4] describes a method called Minimum Error Gamut Compression (MEGC) in which all in-gamut colors are left unchanged and all out-of-gamut colors are mapped to the closest in-gamut color as measured by the euclidean distance in L*a*b* space. There is no differentiation between changes in hue, lightness or chroma. The MEGC method was judged to have produced the best results in terms of closest appearance to the originals when compared with the output from other techniques also considered in his thesis. The MEGC method still has a few significant problems. There is a tendency to lighten the shadows; when this was corrected the dark, saturated colors would be mapped to black. In addition smooth gradients were sometimes reduced to steps and objectionable hue shifts could occur.

ADAPTIVE GAMUT COMPRESSION

Although the above techniques clearly can be made to produce excellent results, either the need for operator skills or unacceptable artifacts and errors makes them unsatisfactory for use in a commercial sense. The goal of the following method is to strike a balance between global scaling and clipping. The enormous variety of potential images also dictates that a superior solution will take into account the particular needs of the individual image. It is undesirable to modify the entire image if the only out-of-gamut colors are a series of saturated reds, for example. Images may contain other problems besides out of gamut colors that will detract from the quality of their reproduction. The method described below is restricted to those images that have no defects such as a color imbalance and are satisfactory to the user apart from the fact that their gamut exceeds the printer gamut. The fundamental premise underlying the technique is that the primary improvement is obtained via a lightness compression which is done by transforming the image data into CIELAB space and filtering the L* component. Adjustments for limitations in the printable saturation levels are made in the chromatic variables using the hue angle and chroma definitions for CIELAB. This decomposition of the problem contrasts with the more typical approach of employing shifts in lightness to offset loss of saturation.

LIGHTNESS COMPRESSION The method is based upon homomorphic image processing techniques. It was assumed that the image lightness data [L*] could be represented in the following form:

$$F(x,y) = I(x,y)*R(x,y)$$

where $F(x,y)$ is the two dimensional image data, $I(x,y)$ the incident illumination and $R(x,y)$ the scene reflectances. In this way the image is decomposed into two components: $I(x,y)$ which is assumed to be a slowly varying function dependent upon the illuminating radiation and $R(x,y)$ as the rapidly changing function which will contain the detail information for the image. $I(x,y)$ controls the dynamic range of the image and under the above assumption can be extracted via filtering in the spatial frequency domain. This is accomplished by taking the logarithm of $F(x,y)$ which is, in principal, $\log I(x,y) + \log R(x,y)$ and low pass filtering the Fourier transform to obtain $I(x,y)$ and $R(x,y)$. This enables the dynamic range to be scaled to fit that of the hard copy output device. This process is directly analagous to a photographic technique in color separation known as area masking which is, in reality, extremely unsharp masking. A 2D gaussian impulse was used as a filter for this stage with a time constant specified in pixels. This was not found to be a sensitive parameter and values between 5 and 20 were found to work quite well. The low frequency image was then scaled and recombined with the high frequency image, similar to the manner in which an area mask would be used. Correction for the overall darkening of the resultant combination was made and the data scaled between the endpoints of the printer's lightness range. The end result was a new L* image whose dynamic range matched that of the printer.

The difference between this approach and the more common one of reducing the lightness range by the application of a static nonlinear transformation to the pixel values [6] is that contrast is allocated as a function of frequency. Reducing contrast as a function of lightness monotonically reduces the dynamic range of the image. For the best results this requires operator intervention based on the properties of the individual image otherwise it is possible to unnecessarily reduce the contrast of the image. Decreasing contrast as a function of frequency does not monotonically lower the dynamic range of the image. In the process described above the filtering process is halted once there is is no accompanying reduction in dynamic range. In this manner the process adapts itself to the particular image and at the same time maintains as much local contrast as possible. In other words, the higher frequency components of the image are sharpened. This approach is justified on the basis that the higher frequency components i.e. the details of an image, are almost always more important to image quality than low frequency components.

CHROMA COMPRESSION The second processing step is designed to smoothly desaturate only those regions of the image that are outside of the printer's color gamut. A Hewlett-Packard PaintJet ink jet printer was used as the hard copy device. The solid volume of printable colors was quite angular and irregular when represented in CIELAB space. Due to the unwanted absorptions of the ink dyes the printer has a hue and lightness dependent performance when it comes to saturated colors, represented by the irregular shape of the gamut solid. Global scaling will not make the best use of the printer gamut since the worst part of the printer saturation gamut will dominate the calculation of the scaling parameter. Clipping the out-of-gamut colors to the boundary will lead to loss of detail and in cases where the image gamut far exceeds the printer gamut, a large number of saturated colors will "pile up" at the boundary producing an unrealistic result. A combination of both techniques is clearly the best approach. The chroma compression algorithm used was also designed to be adaptive to avoid being controlled by that portion of the printer gamut which was most limited in performance.

The amount of compression is a function of hue (h*) and

lightness L* and changes smoothly to avoid discontinuities in the mapping due to the irregular shape of the gamut solid. Hue (h*) is the CIE 1976 a,b hue angle

$$h* = \arctan(b*/a*)$$

and chroma is the CIE 1976 a,b chroma. If all chroma values for a particular (L*,h*) region are inside the color gamut then no compression will be performed on the region. If chroma values are outside of the color gamut then they will be mapped into the gamut and some of the colors that were inside the color gamut will be moved towards the achromatic axis to make space for the out-of-gamut colors. Those colors that are inside the color gamut but altered by the scaling are desaturated in order to avoid a many to one mapping near the gamut surface that can produce false contours. The reduction in chroma is always kept to a minimum to keep the image as colorful as possible.

CLIPPING The purpose of this third and last step is to insure that all pixel values are within the printer's color gamut without forcing either of the previous two steps to overcompress the image. Clipping is applied only after the lightness and compression changes have been made. Hue and lightness are unaltered by this step. Any out-of-gamut color is simply moved to the point on the gamut surface that has the same hue and lightness value.

RESULTS

The subjective nature of the evaluation of the printed result does not reduce to a graphical representation but some overall comments can be made. The lightness compression was found to be very effective, although computationally intensive. A wide range of images from offset litho to photographic prints were scanned and processed. The initial pass through the lightness compression step did produce a pleasing result in all cases without the need for changing any process variables. Chroma compression for offset litho originals was indistinguishable from simple clipping but was essential for photographic prints. The detail sharpening was less than expected and did not match that obtainable from an edge enhancement algorithm. However the intent was not to enhance the image detail but minimize any degradation due to the lightness compression. The interaction of a sharpening algorithm with the above image processing technique was not explored. Further work, including the incorporation of a sharpening algorithm, is required before a final evaluation of this method as an automatic gamut mapping process can be made.

SUMMARY

The subjective nature of color images requires that any scheme for gamut mapping must take into account the content of the image if optimum results are desired. Much of the success of any scheme depends on the lightness scaling which should be performed adaptively as must also the chroma compression. If these two steps are successful then the number of objectionable hue errors will be small and tolerable.

REFERENCES

[1] M. C. Stone, W. B. Cowan and J. C. Beatty, "Color Gamut Mapping and the Printing of Digital Images", Xerox Corporation, April 1988.

[2] M. C. Stone, W. B. Cowan and J. C. Beatty, "A Description of Color Reproduction Methods Used for This Issue of Color Research and Application", Color Research and Application Vol 11, pp S83- S88, supplement 1986.

[3] M. C. Stone, W. B. Cowan and J. C. Beatty, "Color Gamut Mapping and the Printing of Digital Color Images ", ACM Transactions on Graphics, Vol. 7, No. 4, October 1988, Pages 249-292.

[4] R. R. Buckley, Jr., "Reproducing Pictures with Non-reproducible Colors", S.M. Thesis, Massachusetts Institute of Technology, February, 1986.

[5] J. J. Sara, "The Automated Reproduction of Pictures with Non-reproducible Colors", Ph.D. Thesis, Massachusetts Institute of Technology, August 1984.

[6] W. F. Schreiber, "Image Processing for Quality Improvement", Proceedings of the IEEE, Vol. 66, no. 12, Pages 200-211, August 1986.

Challenges in Device-Independent Image Rendering

Robert A. Ulichney

Digital Equipment Corporation
550 King St., LKG1-2/C13, Littleton, MA 01460-1289

Abstract

At display or print time, images must be properly scaled with optional sharpening, tone-scale or color adjusted, and quantized either dynamically or simply, depending on the available color levels at the targeted device.

1. Introduction

The goal of incorporating images as a generic data type (like ASCII text) in general purpose computer systems poses a considerable design challenge. The user has access to wide range of display devices that vary dramatically in resolution and size and color or gray-scale quantization values. It should be a transparent operation to send an image to an arbitrarily shaped workstation window, a terminal with asymmetric pixels, or a bitonal hard copy device, and have the fidelity of the original preserved as best as it can be.

This paper addresses this problem from a systems point of view and presents design tradeoffs that exist. Perhaps the most difficult challenge is that of gamut mapping between devices; to complement the other papers in this session that focus on this topic, this paper will instead concentrate on the other parameters that must be satisfied for device-independent rendering to work.

It should be clarified that in this paper the use of the term "rendering" should not be confused with the sense used in 3-D graphics synthesis.

2. Rendering Model

An overview of a system for image rendering is shown is the figure. The task of the "Render" block is take decompressed image data and tailor it for a specific target display. Successful rendering requires intimate knowledge of the nature of the display device; this can be modeled as the Physical Reconstruction Function [1, ch. 2].

With the soon-to-be-adopted DCT-based international standard for continuous-tone color image compression [2], it is reasonable to assume that source images will be stored so compressed in a "pristine" device-independent way. The acceptance of a standard will hopefully mean readily available hardware to quickly perform decompression.

The rendering system comprises the three stages of (1) Resample/Sharpen, (2) Color/Tone Adjust and (3) Dynamic or Simple Quantization. In the first stage, the original must be resampled to match the grid of the target. Most often this will mean a simple digital enlargement or reduction; however, the possibility exist for a change in pixel aspect ratio, or even a rectangular to hexagonal lattice conversion. For simple rectangular-to-rectangular grid scaling, the best filters to use have been determined from a perceptual point of view [3]. When bandlimiting

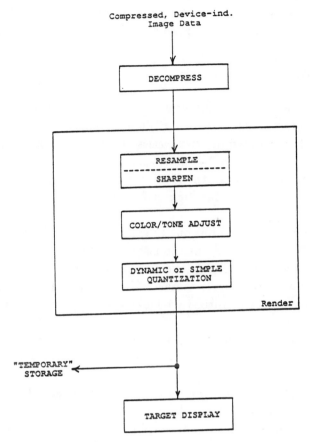

Image rendering system. Parameters are tailored for each target display.

for reduction, a Gaussian with $\sigma = 0.30\times$output-period is recommended. For interpolation, it was found that the filter most preferred in terms of looking the most like the original was a cascade of two: first sharpen with a Laplacian followed by convolution with a Gaussian with $\sigma = 0.375\times$input-period.

A typical sharpening scheme can be expressed as follows:

$$J_{\text{sharp}}[x, y] = J[x, y] - \beta\Psi[x, y] * J[x, y].$$

where $J[x, y]$ is the continuous-tone input image, $\Psi[x, y]$ is a digital Laplacian, and "$*$" is the convolution operator. The nonnegative parameter β controls the degree of sharpness. Besides the need for sharpening in interpolation, it belongs in the rendering model for other reasons. If sharpening is desired on an image that is being reduced in scale, it must occur *after* scaling and thus cannot be done outside of the rendering system. Also, if dithering is later used as the quantization means, it may be needed to compensate for a loss of high spatial frequencies.

Convolutions are expensive. For scaling up or down by factors less than 2.5, nearest neighbor (convolutionless) scaling is usually adequate.

The problem of gamut mapping is handled in the second stage. Again, this difficult topic is addressed by the other papers in this session. For the case of monochromatic displays, a one-dimensional mapping can be established by direct measurement of a test gray ramp.

When sufficient color quantization levels exist so as to avoid spurious contours, a simple uniform quantizer is all that is needed in the third stage of the rendering system. Perhaps most commonly, however, is the case where the targeted device has an insufficient number of colors or frame-buffer memory requiring what can be called "dynamic quantization".

3. Dynamic Quantization

There are three basic classes of techniques for circumventing the problem of insufficient colors or color memory: (1) histogram-based methods, (2) chrominance subsampling, and (3) dithering.

Histogram-based methods all require two passes of the entire image data; the first to acquire the histogram statistics from which a 3-D quantizer to N colors is fabricated, and a second to perform the pixel assignments. Perhaps the fastest method is the popularity algorithm [4] where a simple sort finds the N colors with the highest frequency, and all other colors are mapped to those.

A more compute intensive method but one that in general performs much better is the often used median-cut algorithm [4]. In this method the color space is repeatedly subdivided into smaller rectangular solids at the median planes, the goal being that each of the selected colors represent an equal number of colors in the image. The average of the colors in each of the final regions are the colors used in the quantizer. A later, less compute-intensive variation on this is the mean-split algorithm.

More recently, a clustering technique has been reported [5] that has less quantization error than the above mentioned methods, as minimization of sum-of-squared-errors is central to its design.

Chrominance subsampling [6,7,8] exploits the perceptual fact that chrominance acuity is much less than that for luminance. Typical implementations average each of the two chrominance values in a given luminance-chrominance color representation over either a 2×2 or 4×4 region; this results in an average of 12 or 9 bits per pixel respectively assuming 8 bits of amplitude resolution per component. This approach requires hardware, usually at video rates, to up-sample the chrominance components and convert the color space. The tradeoff is special purpose hardware for less frame-buffer memory.

The third alternative for dynamic quantization is a dithering method. Several methods exist [9,10,11,1], primarily designed for binary output but all extendible to multilevel color. The basic principle is to use the available subset of colors to produce the illusion of any color in between by judicious arrangement.

The computationally simplest method is the point process of ordered dither where a deterministic "noise" array tiles the plane in a periodic manner. Each component of each pixel in the image has associated with it a "noise" or dither amplitude that is added to it before being passed to a uniform quantizer. Images on most electronic devices look best with what is called a "dispersed-dot" dither pattern, patterns with take best advantage of the available spatial resolution to produce homogeneous distributions of noise. Some bitonal devices cannot accommodate isolated dots so a clustered-dot such as the printer's screen works best in that case.

Because the patterns are periodic, ordered dither does suffer from the low frequency texture associated with the periodicity. Where additional computation can be afforded the aperiodic operation of error diffusion-based techniques produce pleasing results. Error diffusion uses a uniform quantizer but distributes the quantization error for each pixel over some neighborhood of yet-to-be-quantized pixels.

Combinations of these quantization alternatives can be used but the increase in computation begins to become less fruitful in terms of quality increase.

4. Architectural Concerns

The default parameters for such a rendering system should be chosen so as to best preserve the integrity of the source image. When the user needs to change the image from these defaults, controls should be supplied for, say, tint- and luminance-control, sharpness, and scale changes. But what processing time per change should the system be designed for? (Speed is not that critical for hard-copy rendering, but is in the workstation environment.) Is .5 seconds necessary or can 10 seconds be tolerated? Such requirements have enormous cost consequences as the choice is between expensive special purpose hardware or a general purpose processor-based software solution.

If rendering speeds are inhibitive for certain applications where the same images are repeatedly displayed on the same device, it is important to allow the device-dependent rendered image to be saved in what is called "temporary" storage in the figure.

Making an image data type easy to use depends on a workable device-independent rendering strategy; this is perhaps the next fundamental challenge in computer system design.

References

[1] Ulichney, R.A., *Digital Halftoning*, Cambridge: The MIT Press (1987).

[2] Hudson, G.P., H. Yasuda and I. Sebestyen, "The international standardization of a still picture compression technique," IEEE Global Telecomm. Conf., pp. 1016–1021, Nov 28–Dec 1 (1988).

[3] Schreiber, W.F. and D.E. Troxel, "Transformation between continuous and discrete representation of images: a perceptual approach," *IEEE Trans. PAMI*, vol. PAMI-7, no. 2, pp. 178–186, (1985).

[4] Heckbert, P.S., "Color image quantization for frame buffer display," *Computer Graphics* (AMC SIGGRPAPH'82 Conf. Proc.), vol. 16, no. 3, pp. 297–307 (1982).

[5] Wan, S.J., K.M. Wong, and P. Prusinkiewicz, "An algorithm for multidimensional data clustering," *ACM Trans. on Math. Software*, vol. 14, no. 2, pp. 153–162 (1988).

[6] Sigel, C., R. Abruzzi, J. Munson, "Chromatic Subsampling for display of color images," —*this conference*— (1989).

[7] Luther, Arch, *Digital Video in the PC environment*, New York: McGraw-Hill, pp. 193–194 (1989).

[8] Glass, L.B., "Digital Video Interactive," *Byte*, May, p. 284 (1989).

[9] Roetling, P.G., "Binary approximation of continuous-tone images," *Photographic Science and Engineering*, vol. 21, pp. 60–65 (1977).

[10] Stoffel, J.C. and J.F. Moreland, "A survey of electronic techniques for pictorial reproduction," *IEEE Tran. Commun.*, vol. 29, 1898–1925 (1981).

[11] Jarvis, J.F., C.N. Judice, and W.H. Ninke, "A survey of techniques for the display of continuous-tone pictures on bilevel displays," *Computer Graphics and Image Processing*, vol. 5, pp. 13–40 (1976).

Color Rendering from CMYK Inks into Film Dyes

Jay Thornton, Dan Bybell, Bill Donovan and John McCann
Polaroid Vision Research Laboratory
750 Main Street 3E   Cambridge, MA.   02139

It is often difficult to obtain quality hard copy for high resolution (eg. 6000x3000 pixels) digital images. To obtain hard copy, we use a film recorder which writes a fine image raster directly onto Polaroid instant color print film. Our concern that the color in the print be controlled as precisely as possible led to the development of a special purpose color computer. The color computer resides in the film recorder and transforms the input image pixel values into those output exposure control values necessary to attain the desired color on film. This color transformation is done on the fly without slowing down the printing of the image.

The film recorder (the Instant FIRE 300 manufactured by MacDonald Dettwiler Technologies Ltd.) images the raster on film as follows (see figure): a vacuum holds the film negative against the inside of a semi-cylinder. Light for exposure originates from a Xenon arc; is filtered to produce red, green, or blue; and finally is collimated to travel down the center of the cylinder. On the axis of the cylinder is a lens and a mirror which focus the light onto the film. The mirror rotates at 60 Hz. writing one color of one line each revolution. After all three colors have been written, the entire mirror assembly steps along the cylinder axis and repeats the process at the next line. The exposure is controlled by acousto-optic modulators (AOMs). By synchronizing the light modulation to the spinning mirror, the exposure of individual pixels on each line can be controlled. We typically write about 320 pixels/inch in each direction, and have an 8-bit exposure scale in each color. It takes about 12 minutes to expose a 12" by 18" image and another 2 minutes for processing.

The film recorder was developed for the printing industry. Printers frequently deal with 100 Mbyte images which have been digitally manipulated (eg. to remove blemishes from faces or entire objects from a scene). Before going to the expense of actually printing the image, one would like to check for errors by creating a pre-press proof which accurately renders what that image will look like as ink on paper. Making such a proof on film requires writing an image which is pixel by pixel metameric to an image produced by a printing press. Our special purpose hardware to perform a color transformation from the cyan, magenta, yellow, and black representation used in 4-color printing into the red, green, and blue exposure values needed to achieve metameric colors on film. Calculating such a transformation requires models of the color printing process, models of color film, and a method of linking the models.

For many printing applications, the Neugebauer model does a good job of predicting the composite color obtained for an arbitrary CMYK combination. This model assumes that the four basic inks overlap to produce $2^4=16$ possible colors whose color can be directly measured. The total color is a weighted sum of the 16 possible colors in which each color is weighted by the fractional proportion of surface area it occupies. The fractional areas of the 16 colors are calculated from the fractional amounts of the four basic inks by making an assumption similar to statistical independence (eg. the fraction of green is equal to the fraction that is cyan multiplied by the fraction that is yellow.) Other printing technologies, like gravure where the inks are not confined into sharp dots, required that we develop new models for the predicition of ink colors.

One factor complicating the linking of film and ink colors is the difference in the color gamuts available. Ink on paper is capable of reproducing a lighter white and a more saturated yellow than photographic film. The differences in the color gamuts had to be included in the process of establishing a corresponding film color for each possible ink color.

The linking of ink and film colors is further complicated by factors not included in the common understanding of the term "metameric". For instance, the difference in the surfaces of the two media--film being glossy while the ink on paper is matte--means that an appearance match will be dependent on the viewing geometry. The same problem arises with respect to the instrumentation with which color is quantified in the laboratory. Neither of the common measurement geometries (integrating sphere or 45°/0°) adequately capture the psychophysical matches made in typical, real-world viewing geometries.

Successful pre-press proofs require that the color transformation be capable of high precision so as to have a close match between film and press sheet. Fundamental to this high precision is the ability to attain a desired film color on demand. This is possible through a brute force calibration of a large subset of the film gamut. Fitting a film model to these data permits smooth interpolation between sample points and helps suppress noise present in the calibration data. This model ultimately produces a colorimetric characterization of the film response to an arbitrary exposure. After loading this colorimetric information into the color transform hardware, one can specify colors to be written to film (say in L*a*b* units) and the appropriate exposure will be employed to attain that color. Expected errors in the L*a*b* coordinate system are on the order of 1 unit.

We have demonstrated the precision and generality of the color transform hardware and software by producing "pseudo-isochromatic plates" for the detection of color blindness. The creation of these plates was made simple by the use of a color transformation which was calculated to connect variation in the first band of a three band image to variation in the quantum catch of the long wavelength sensitive photoreceptors only. The second band was connected to variation in the middle wavelength sensitive receptors, and the third to the short. An individual missing one of the photoreceptor classes (a "dichromat") will be unable to see a messages written in the image band corresponding to the missing receptor class.

DIRECT DIGITAL FILM RECORDER

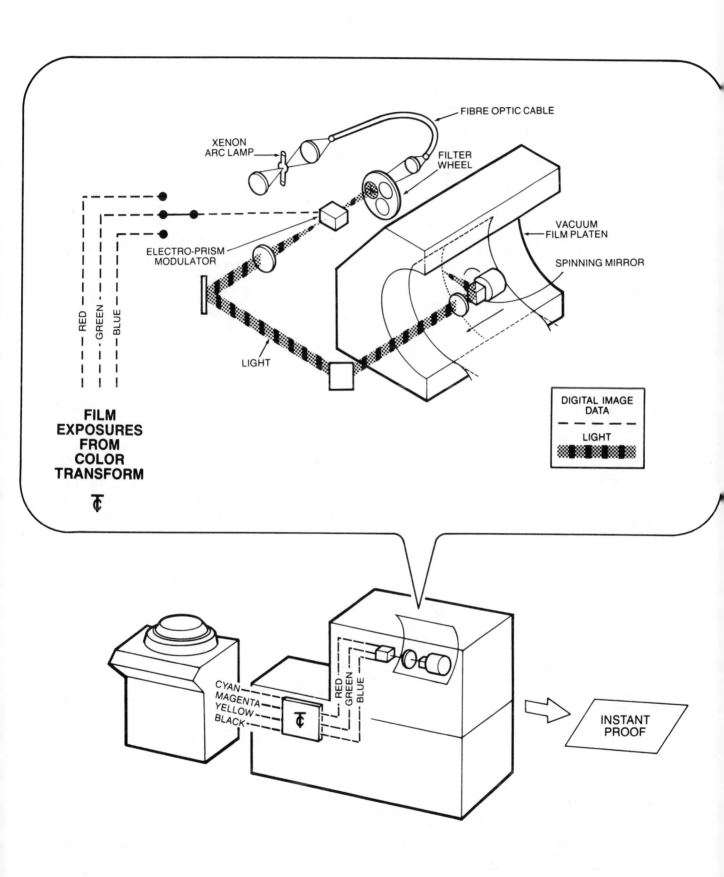

FIBRE OPTIC CABLE

XENON ARC LAMP

FILTER WHEEL

VACUUM FILM PLATEN

SPINNING MIRROR

ELECTRO-PRISM MODULATOR

RED GREEN BLUE

LIGHT

FILM EXPOSURES FROM COLOR TRANSFORM

DIGITAL IMAGE DATA
LIGHT

CYAN MAGENTA YELLOW BLACK

RED GREEN BLUE

INSTANT PROOF

EMPIRICAL STUDIES OF COLOR MATRIX DISPLAY IMAGE QUALITY

Louis D. Silverstein, Frank E. Gomer, Yei-Yu Yeh, and John H. Krantz
Systems and Research Center
Honeywell Inc., Phoenix, Arizona 85036

Introduction

Color matrix display (CMD) technology has evolved to the point of viability for many information display applications. The matrix-addressed color liquid crystal display (LCD) currently appears to be the most suitable CMD technology for producing full-color images. Relative to color displays based on the shadow-mask cathode ray tube (CRT), the benchmark technology against which all other color display technologies must be evaluated, CMD panels offer potential improvements in design flexibility from the standpoint of relatively low power requirements, smaller volume, increased reliability, and better image visibility under high-ambient lighting conditions. These attributes make the CMD particularly attractive for vehicular and field-based display applications.

A significant limitation of the present generation of CMDs is the relatively poor image quality compared to that found with CRTs. The quality of the CMD image may be described by the degree to which two distinct types of image artifacts, chromatic and spatial distortion, are visually detectable in the image. Chromatic distortion, which is characterized by color banding, color fringing, and detectable individual primary color elements in mixture colors, results from the failure of the human visual system to spatially integrate individual primary color elements or patterns of elements into an image of the desired color. Spatial distortion, which is perceived as stairstepping in lines and arcs, jagged edges, and gaps or luminance bands in primitive graphics and alphanumeric characters, results from the interaction of several imaging parameters. These parameters include the resolution or sampling density of the CMD, image quantization noise produced by individual discrete elements and the geometric pattern or mosaic into which the elements are arranged, truncation errors due to the limited number of available gray levels, and the bandwidth requirements of the original graphic image.

Fortunately, CMD technology does offer the potential for excellent image quality which may ultimately exceed that obtainable with the high-resolution color CRT. The source of this imaging potential results from several important attributes of the LCD-based CMD: 1) flexibility in the definition of the color image sampling mosaic, including size, shape, spacing, and luminance profile of elements; 2) addressability and control of every imaging element; 3) decoupling of the image-forming function from the light-generating function; 4) relative insensitivity to the ambient lighting environment; and 5) stability and homogeneity of chromaticity and luminance across the imaging surface.

While CMD technology provides significant opportunities for display system design, it poses great challenges for applied vision scientists and human factors specialists. Display visual parameters must be carefully specified. Design trade-offs must be assessed and evaluated in light of fabrication and production costs. Moreover, imaging performance resulting from any given set of design parameters must be clearly demonstrable and, to the extent possible, quantitatively characterized and scaled relative to system development costs.

As part of a continuing research program to investigate CMD imaging characteristics and improve the image quality of this important new generation of display devices, we have previously developed a high-fidelity CMD image simulation system,[1,2] established and validated a sensitive psychophysical methodology for direct scaling of image quality judgments,[3,4] empirically evaluated addressing algorithms and color pixel mosaics for CMDs with binary luminance states,[3,5] and investigated pixel mosaics, anti-aliasing approaches, and gray-level requirements for CMDs with gray-scale capability.[5,6] The present paper briefly describes our CMD image simulation system and experiments which address three critical issues in CMD design: 1) preferred pixel mosaics for binary CMDs used in graphics applications; 2) preferred pixel mosaics for CMDs with gray-scale capability; and 3) the relationship between the number of gray levels and CMD image quality for graphic images.

The Honeywell CMD Image Simulation System

The Honeywell CMD Image Simulation System[1,2] enables the generation of high-fidelity visual emulations of CMD images through accurate control of: a) pixel size, shape, and separation; b) pixel chromaticity; c) pixel luminance and luminance profile; d) pixel mosaic; e) gray scale; f) pixel addressing algorithm; and g) image viewing distance. The system consists of color graphics image processing hardware and software, as well as high-resolution display and optical components. Simulated CMD images are produced by initially generating the images in magnified form on a 48.3 cm diagonal high-resolution (1280 x 1024) color monitor. The magnified images are then passed through a binocular pair of optical channels which are used to create a minified virtual image of the screen. Virtual image distance (i.e., system focus) and binocular disparity are set within the system so as to provide appropriate visual accommodation and stereoscopic cues for any selected viewing distance. Figure 1 shows the hardware configuration of the image simulation system.

Figure 2 illustrates the process by which a simulated yellow CMD image, for example, is generated on the observer's color monitor. The top panel (A) of Figure 2 shows the complete simulated image. This image is several times the size of an actual CMD image and is viewed through the reduction optics of the simulation system. The image is composed of individual, simulated, primary color CMD pixels which are arranged according to a predefined mosaic (panel B). In turn, each simulated CMD pixel is composed of many smaller "sub-pixels," which actually correspond to individual triads of red, green, and blue phosphor dots on the faceplate of the monitor (panel C). By using a relatively large number of sub-pixels to create the simulated CMD pixels, several important spatial attributes of actual CMD pixels can be retained. These include pixel size and shape, inter-pixel spacing, and the luminance distribution across the pixel surface.

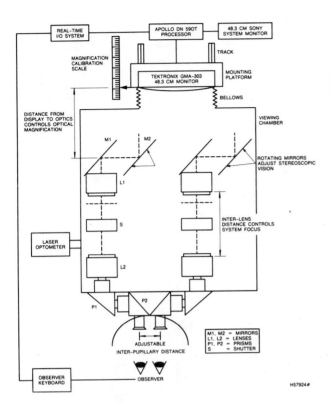

Figure 1. Hardware configuration of the CMD image simulation system

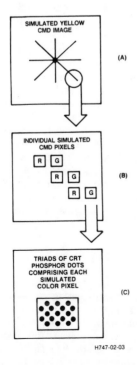

Figure 2. Process by which a simulated CMD image is generated on the high-resolution monitor.

General Experimental Methods

The experiments were conducted using the CMD image simulation system described above. For the present studies, the system was calibrated to produce 0.16 mm x 0.16 mm square pixels with an inter-pixel spacing of 0.021 mm as viewed through the system optics with a 6:1 optical minification ratio and a virtual image distance of 81.28 mm. The display was divided into four 1.5° square quadrants, each of which contained a completely independent simulated CMD image defined by a unique set of attributes. Six test colors [red (R), green (G), yellow (Y), cyan (C), magenta (M), and white (W)] and three different color pixel mosaics (RGB diagonal, RGB delta triad, and RGBG quad with diagonally opposed green elements) were evaluated in the experiments. The chromaticity coordinates and peak luminance values for the simulated pixels/primary color test images and the secondary color test images are shown in Table 1.

Table 1

1931 CIE Coordinates and Luminances

Simulated Pixels/Primary Color Test Images			
Color	x	y	Luminance (cd/m^2)
Red	.5930	.3540	25.70
Green	.2782	.6077	52.73
Blue	.1740	.1522	15.38
Green (RGBG)	.2782	.6077	26.37
Secondary Color Test Images			
Color	x	y	Luminance (cd/m^2)
Yellow	.4216	.4921	78.43
Cyan	.2221	.3626	68.11
Magenta	.3492	.2366	41.08
White	.3255	.3602	93.81

A magnitude estimation procedure was used to quantify observers' psychophysical judgments of CMD image quality.[3,4] Separate magnitude estimates were obtained for judgments of spatial quality, color quality, and overall image quality of the test images. A standard reference image, written in high-resolution raster, was always displayed in the lower left quadrant of a test frame and served to anchor observers' magnitude estimates to a maximum value of 100 (i.e., to an "ideal" image). Prior to an experimental session, participants were given extensive practice on the magnitude estimation procedure and detailed instructions on the types of image features which were important to each type of judgment. Twelve observers participated in each experiment and were screened for normal or corrected-to-normal vision for the following functions: near and far visual acuity; near and far vertical and lateral phoria; stereopsis; and color vision.

Except where noted, all magnitude estimation judgments were converted to range-corrected scores. This transformation compensates for the fact that different observers tend to use different number ranges when reporting their judgments. Since both experiments utilized complete factorial within-subjects experimental designs, statistical analyses of the data were based upon repeated-measures analyses of variance using the Greenhouse-Geisser conservative F test with reduced degrees of freedom. Post hoc comparisons were performed with Tukey's

Honestly Significant Difference (HSD) Test and also used conservative, reduced degrees of freedom. All reported effects are statistically significant at p < .05 or better.

Due to the brief nature of the present paper, only the results for magnitude estimates of overall image quality are described below. While the results for spatial and chromatic judgments are revealing of specific image quality components, the composite magnitude estimates of overall image quality provide the most coherent impact on CMD systems design.

Experiment 1

In the first experiment, we focused on the simultaneous visual comparison of the three pixel mosaics (RGB diagonal, RGB delta triad, and RGBG quad) used in a binary mode of graphic image generation. A "sliding window" pixel selection algorithm, established by a previous study as a superior approach for addressing the three pixel mosaics, was used to generate the test images.[3] Basically, this algorithm fixes an aperture or window about the centroid of an intended line segment. All simulated pixels of the appropriate color falling within the window (a 50% area criterion was adopted for pixel selection) are illuminated. The absolute window width varies dynamically, depending upon the pixel mosaic being addressed and the angle of the intended line segment. For any angular orientation, the algorithm defines the smallest possible addressing window for constructing color balanced lines, consistent with the objectives of maximizing effective display resolution and minimizing color errors[4]. The average window width, across the three pixel mosaics and 360° line rotation, was approximately 2.5 pixels.

Three types of test images were used: a static, star-like pattern composed of five intersecting line segments whose angular orientations were selected to represent worst-case sampling axes for each of the pixel mosaics; a static, concentric circle pattern composed of two circles; and a single, straight-line segment which rotated continuously through 360° at a rate of 6° per second. Four images of the same type and color were presented simultaneously in the four quadrants of each display frame. Pixel mosaic was varied within a display frame, while image type and color were varied between display frames. Pixel mosaics were randomly assigned to three of the display quadrants (the "standard" was always presented in the lower left corner), with the constraint that across the experiment each pixel mosaic appeared equally often in each display quadrant. The ordering of test conditions for image type and color was counterbalanced across the 12 observers. Each pixel mosaic and test image combination was tested twice for each of the six display colors. This yielded a 3 x 3 x 6 x 2 within-subjects, factorial experimental design, representing the factors of pixel mosaic, test image type, color, and replication, respectively.

Results

The results from the first experiment revealed highly significant main effects for pixel mosaic, image type, and image color. Post hoc analyses for the pixel mosaic main effect indicated that the RGBG mosaic was judged to produce images with significantly better overall quality than either the RGB delta triad or RGB diagonal mosaics. The latter two mosaics did not differ significantly from one another. Moreover, as shown in Figure 3, the superior performance of the RGBG mosaic was not found to be specific to the type of graphic image tested. Turning to the main

effect of image type, the concentric circles yielded better overall image quality than either the static line pattern or rotating line. Image quality of the static line pattern and rotating line was judged to be equivalent. Finally, the effect of color on overall image quality was manifest in three clusters of colors: white was found to be significantly better than all other colors; the secondary colors yellow, cyan, and magenta did not differ from one another but were all significantly better than either red or green; and the primary colors red and green consistently yielded the poorest overall image quality.

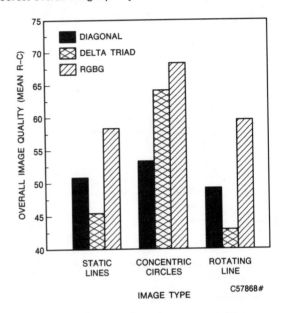

Figure 3. Mean magnitude estimates of overall image quality as a function of pixel mosaic and type of image

The analyses of results from Experiment 1 also revealed a significant interaction between pixel mosaic and image color. Post hoc comparisons indicated that this interaction was attributable to a consistent pattern of results. For the colors white, yellow, green, and cyan, the RGBG mosaic yielded significantly better overall image quality than either the RGB delta triad or RGB diagonal mosaics, with the latter two mosaics providing equivalent image quality (except in the case of cyan, where magnitude estimates were higher for the RGB delta triad mosaic). However, for the colors red and magenta, the RGB delta triad and RGB diagonal mosaics were comparable and both produced significantly better overall image quality than the RGBG mosaic. The pixel mosaic by image color interaction is depicted graphically in Figure 4.

Experiment 2

The second image quality experiment investigated gray scale requirements for anti-aliasing or bandwidth limiting of graphic images presented on CMDs. Gray levels were drawn from an equal partitioning of the linear luminance ramps for each of the simulated primary color pixels. The lowest gray level was always set to equal zero (i.e., the luminance of the display background), while the highest gray level was always set to the maximum luminance for a primary color pixel. Gray levels were used to construct a best-fit sampled approximation to a Gaussian line spread function (LSF) using sub-pixel resolution to determine gray level assignments. Both the Gaussian LSF and linear luminance ramp were selected on the basis of pre-tests which revealed that: a) the Gaussian profile provided better bandwidth limiting and reduction of sampling noise than several other potential profiles (trapezoidal, triangular, or rectangular); and 2) a linear luminance ramp was superior to either logarithmic or power functions. The half-amplitude width of the LSF was approximately 0.508 mm.

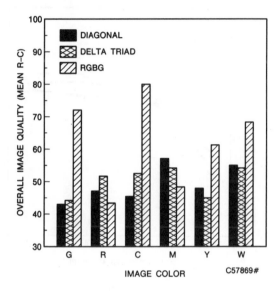

Figure 4. Mean magnitude estimates of overall image quality as a function of image color and pixel mosaic (binary CMDs).

Three gray scales (4, 8, and 16 levels) were factorially combined with three pixel mosaics (RGB diagonal, RGB delta triad, and RGBG quad), six display colors, and two types of test images (static, star-like line patterns and rotating line segments). Four test images of the same type, color, and gray scale were presented simultaneously in the four quadrants of each display frame. Pixel mosaic was varied within a display frame, while gray scale, image type, and color were varied between display frames. Pixel mosaics were randomly assigned to display quadrants, with the constraint that across the experiment each pixel mosaic appeared equally often in each display quadrant. The ordering of test conditions for gray scale, image type, and color was counterbalanced across the 12 observers. Each combination of factors was tested twice, resulting in a 3 x 3 x 6 x 2 x 2 within-subjects, factorial experimental design representing the factors of pixel mosaic, gray scale, color, test image type, and replication, respectively.

Results

Magnitude estimates of overall image quality were significantly influenced by pixel mosaic, gray scale, and color. Post hoc analyses of the pixel mosaic main effect revealed that the RGB delta triad mosaic produced images with significantly better overall quality than either the RGBG or RGB diagonal mosaics. Furthermore, overall image quality was significantly better for the RGBG mosaic than for the RGB diagonal mosaic. Regarding gray scale, overall image quality was judged significantly better when test images were constructed with eight and sixteen levels of gray rather than with four levels of gray. No statistically signif-

icant difference was observed between images constructed with eight and sixteen levels of gray, indicating that for simple graphic images presented on CMDs the effect of gray scale for improving CMD image quality asymptotes somewhere between four and eight gray levels. Post hoc analyses of the color main effect revealed that white images had the best overall image quality and were significantly better than red test images. Also cyan test images were significantly better than red test images. In terms of the ordering of colors, the general pattern was similar to that found in the first experiment, with white, secondary colors, and primary colors clustered in three categories of decreasing overall image quality.

The interaction between gray scale and pixel mosaic was found to be statistically significant and is illustrated in Figure 5. A breakdown of this interaction revealed that it was primarily due to the highly significant main effects of gray scale and pixel mosaic. Inspection of Figure 5 corroborates the strength of the main effects and their relative independence.

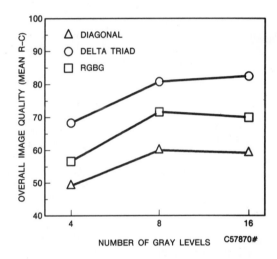

Figure 5. Mean magnitude estimates of overall image quality as a function of number of gray levels of pixel mosaic

The color by pixel mosaic interaction, shown in Figure 6, was also found to be significant. Further analysis of this interaction revealed that the delta triad mosaic produced superior overall image quality for white, yellow, magenta, and red test images. The RGBG mosaic produced significantly better green test images than the other two mosaics, and for cyan images, overall image quality judgments did not differ significantly among the three pixel mosaics. Thus, for four of the display colors (R, Y, M, and W), overall image quality was better with an RGB delta triad mosaic than with either of the other two pixel arrangements.

The final interaction to achieve statistical significance was between pixel mosaic and type of test image. Post hoc analyses indicated that this interaction reflected the fact that rotating test images presented with the RGBG mosaic had significantly better overall image quality than did static images presented with the same mosaic. No significant differences were observed between rotating and static test images presented on either RGB delta triad or diagonal pixel mosaics.

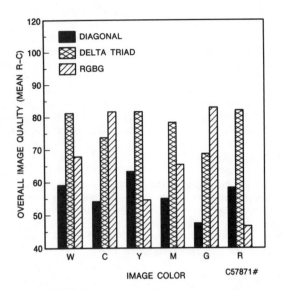

Figure 6. Mean magnitude estimates of overall image quality as a function of image color and pixel mosaic (gray-scale CMDs)

as illustrated in Figure 9, the basic functional relation between image quality and number of gray levels is quite similar for each of the image colors tested.

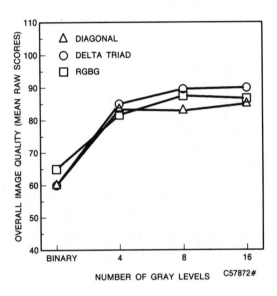

Figure 7 shows the raw (i.e., not range-corrected) magnitude estimates of overall image quality for a simulated binary CMD obtained from the first experiment and for the simulated gray-scale CMD images used in the second experiment. The data for both types of displays are depicted as a function of pixel mosaic, and the data for the gray-scale CMD images are further shown as a function of the number of gray levels used to anti-alias the images. This figure reveals the major improvements in image quality provided by gray-scale anti-aliasing of CMD graphic images. Moreover, the data indicate that for the pixel density and mosaics tested, significant enhancement of CMD image quality occurs only up to 8 levels of gray (3 bits).

Since it is currently difficult to implement stable gray levels in LCD-based CMD technology, two follow-on experiments were conducted to determine the precise point where CMD image quality reaches an asymptote as the number of gray levels is increased. The apparatus, stimulus parameters, and basic configuration of test images were the same as in the main experiment. However, only two pixel mosaics were tested (RGB delta triad and RGBG) and a rank-order procedure was used instead of magnitude estimation. Four test images of the same type and color were presented simultaneously in the four quadrants of each display frame. Number of gray levels was varied within a display frame, while pixel mosaic, image type, and color were varied between display frames. In the first follow-on experiment, image quality was rank ordered for test images with 2 (binary), 4, 6, and 8 gray levels, while in the second follow-on experiment test images with 7, 8, 9, and 10 gray levels were rank ordered. The results from the two follow-on experiments were combined and transformed into an equal-interval approximation based on standard Z-scores.[6]

Figure 8 shows the function relating transformed ranks of image quality to number of gray levels for both the RGB delta triad and RGBG pixel mosaics. From the figure, it is clear that image quality reaches an asymptote at 8 gray levels, independent of the pixel mosaic used to construct the images. Moreover,

Figure 7. A comparison of the overall image quality for binary and gray-scale CMDs as a fucntion of the number of gray levels and pixel mosaic.

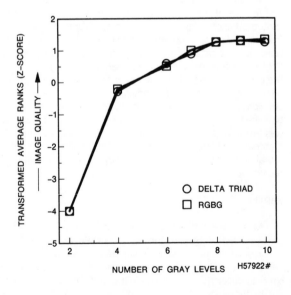

Figure 8. Transformed average ranks of overall image quality as a function of number of gray levels and pixel mosaic

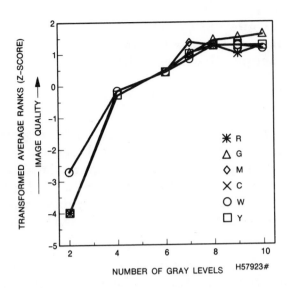

Figure 9. Transformed average ranks of overall image quality as a function of number of gray levels and image color

Conclusions

The results for the first experiment revealed that the RGBG quad pixel mosaic rendered bi-level graphic images of significantly better image quality than either the RGB diagonal or RGB delta triad pixel mosaics, at least for the sampling density and addressing algorithm tested. Further, the superior imaging performance of the RGBG mosaic was manifest for four of the six colors tested (W, Y, C, and G), all of which contain a green component. The results of the first experiment can be attributed to differences in the relative spatial sampling densities and symmetries of the pixel mosaics tested.

Results from the second experiment on CMD gray-scale requirements yielded somewhat different findings. When gray scale was used to bandwidth limit or anti-alias graphically generated lines by imposing a best-fit Gaussian profile across the line, the RGB delta triad pixel mosaic yielded the best overall image quality, regardless of the number of gray levels used to approximate the profile. The significantly superior gray-scale imaging performance of the RGB delta triad mosaic was consistent across both static and rotating line images and, surprisingly, given the results of Experiment 1, was evident for four (R, Y, M, and W) of the six display colors. The incorporation of gray scale provided major improvements in image quality for all three pixel mosaics. Moreover, the data from the main experiment and two follow-on studies revealed that for the pixel densities and mosaics tested, gray scale significantly enhances CMD image quality only up to 8 gray levels or 3 bits. Asymptotic imaging performance is reached at exactly 8 gray levels.

The basic results of Experiment 2 are in accord with previous image quality research using alphanumeric test images[7,8] and are readily interpretable within the context of sampling theory. However, an explanation of the unexpected relative ordering of colors for the RGBG mosaic when gray-scale was employed (attributable to a perceptible low-frequency, luminance speckle pattern for the colors white and yellow which matches the distribution of red pixels) must appeal to wavelength-specific process-

ing mechanisms within the human visual system and appears to reflect the phenomenon of "bleeding".[9] Bleeding occurs when a pattern is too fine to be resolved by the low-acuity chromatic visual system but not too fine for the higher-resolution luminance or form system.

The present paper has described an empirical, methodological approach for addressing visual and perceptual issues in display technology. The results from a series of experiments on CMD image quality have demonstrated the role which applied vision studies can play in solving complex visual display problems.

References

(1) Monty, R. W., Silverstein, L. D., Frost, K, and Boyle, L. (1987). A color matrix display image simulation system for human factors research. Society for Information Display Digest of Technical Papers, pp. 118-122.

(2) Silverstein, L. D., Monty, R. W., Huff, J. W., and Frost, K. L. (1987). Image quality and visual simulation of color matrix displays. Proceedings of the Society of Automotive Engineers Aerospace Technology Conference, Paper No. 871789.

(3) Gomer, F. E., Silverstein, L. D., Monty, R. W., Huff, J. W., and Johnson, M. J. (1988). A perceptual basis for comparing pixel selection algorithms for binary color matrix displays. Society for Information Display Digest of Technical Papers, pp. 156-159.

(4) Silverstein, L. D., Gomer, F. E., Monty, R. W., and Yeh, Y. (1988). An Optimal Pixel Arrangement for Graphic Images Presented on Binary Color Matrix Displays. Honeywell Technical Document No. R74-2891-01-00.

(5) Silverstein, L. D., Monty, R. W., Gomer, F. E., and Yeh, Y. (1989). A Psychophysical Evaluation of Pixel Mosaics and Gray-Scale Requirements for Color Matrix Displays. Society for Information Display Digest of Technical Papers, in press.

(6) Krantz, J. H., and Silverstein, L. D. The Gray-Scale Asymptote for Anti-Aliasing Graphic Images on Color Matrix Displays. (1989). Society for Information Display Digest of Technical Papers, in press.

(7) Rogowitz, B. E. (1988). The psychophysics of spatial sampling. Proceedings of the SPIE/SPSE Conference on Electronic Imaging and Devices.

(8) Cushman, W. H., and Miller, R. L. (1988). Resolution and gray-scale requirements for the display of legible alphanumeric characters. Society for Information Display Digest of Technical Papers, pp. 432-434.

(9) Livingstone, M., and Hubel, D. (1988). Segregation of form, color, movement, and depth: Anatomy, physiology, and perception. Science, 240, pp. 740-749.

Chromatic Subsampling for Display of Color Images

Claude Sigel, RuthAnn Abruzzi, and James Munson

Digital Equipment Corp., Albuquerque, New Mexico 87109

Introduction

The amount of information required to store color images is immense. For a typical 1024x1024 pixel color image, 8 bits each of R, G, and B data are usually stored (to avoid luminance or chromatic contouring artifacts), which adds up to 3 Mbytes per picture. This large size is problematic in several ways: framebuffer memory is still expensive; large framebuffers are technically more difficult to engineer (more boards, more heat); disk storage capabilities become swamped; the time required to transmit a picture from disk to terminal, or between network sites, is unacceptable.

One way to reduce the amount of stored information is to take advantage of any limitations of the human visual system: since it is people who ultimately view the images, any information that will ultimately be ignored by the observer's visual system can be safely discarded from the computer version. Watson [1987] uses this principle in his proposal to efficiently store achromatic visual data. The present study uses this principle to reduce the amount of required color image data by exploiting the visual system's relative insensitivity to chromatic stimuli of high spatial frequency [Granger and Heurtley, 1973; DeValois and Switkes, 1983]. Chromatic information apparently can be *subsampled*, i.e., retained at fewer spatial locations than achromatic information, without introducing perceptual artifacts. (The reduced bandwidth requirements of the NTSC broadcast TV standards [Benson 1986] are also based on this phenomenon.)

Subsampling chromatic, but not achromatic, data cannot be done directly on an image stored as R, G, and B values, because each of those dimensions contain both chromatic and luminance information. Our subsampling algorithm therefore performs an initial transformation of the color data from RGB to a color space that separates chromatic from achromatic dimensions, corresponding to the human opponent processes [Hurvich and Jameson, 1955; Larimer, Krantz, and Cicerone, 1974]. Quantizing the subsampled data results in 8-bit chromatic values that are easily stored in a computer. To display the image on an RGB monitor, the subsampled image data must be interpolated, quantized, and transformed back into RGB space.

Our subsampling algorithm averages and quantizes chromatic data, so the results of this process depend on how the chromatic values are defined. We used two color spaces in this investigation. The YUV space [CCIR, 1983] is a linear combination of RGB, and its chromatic directions are approximately aligned with perceptual opponent processes: positive and negative U and V are red, green, blue, and yellow, respectively. The CIELUV space [Wyszecki and Stiles, 1982, pp 164-169] is more perceptually uniform, and also has approximately red/green and blue/yellow chromatic dimensions, but the transformation from RGB to CIELUV is nonlinear.

Procedures

Figure 1 is a schematic pixel representation of an image. The dots represent pixel locations for which the original eight bits of R, G, and B would be stored in a conventional framebuffer. When the data are converted into one of the alternate color spaces (YUV or CIELUV) there are one achromatic and two chromatic values defined for each pixel. After subsampling, however, chromatic values are retained at fewer spatial locations, represented by circled points in the figure. We define a subsampling spacing parameter, n, which represents the vertical and horizontal spacing of the stored chromatic data. The value of n was varied in the experiments from 1 to 8, and has the value 4 in the figure.

The chromatic values stored at each circled point (circled point 1, e.g.) were calculated by taking a 2-dimensional convolution of the chromatic values surrounding that point (i.e., inside the square outline) with a suitable kernel function. We used a 2-dimensional Gaussian for our kernel, normalized and sized so that the kernel values fall to 1% of their maximum at the edges of the square (for which the standard deviation was defined to be n/2.5).

Figure 1: Pixel representation of an image, showing relative locations of achromatic (dots) and chromatic (circles) information. The subsampling spacing parameter is n = 4.

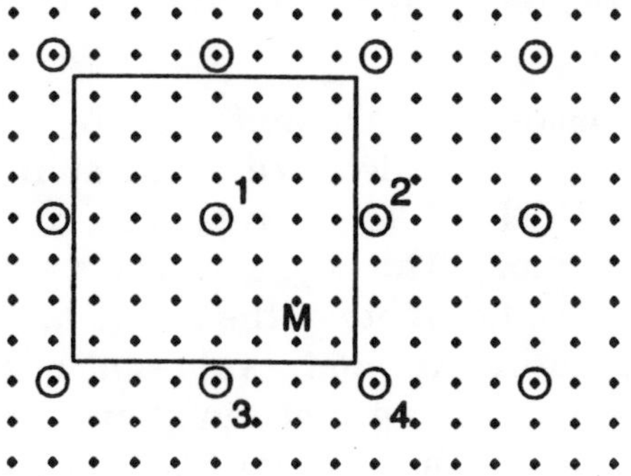

Converting the subsampled data back to displayable RGB values is accomplished by first estimating the chromatic values at every pixel, and then converting the estimates back into RGB values, using the appropriate inverse transform. Chromatic values are estimated with a bilinear interpolation of the four nearest neighbor chromatic values (at circled points 1, 2, 3, and 4, for example, to estimate the values at point M).

The images were displayed on a Sony GDM 1950 monitor driven by a 24-plane Parallax framebuffer. The monitor was calibrated to have a maximum luminance of 100±1 nits, and to have 100 dots/inch both vertically and horizontally. Its gamma was measured and compensated for in hardware lookup tables, and it was viewed from approximately 50 cm.

Eleven observers were used, the three authors and eight who were naive about the aims of the experiment. All had served in similar image assessment experiments at least once before, and were familiar with the particular pictures used in the study. Each observer participated in four sessions, each one of which was concerned with a single 24-plane RGB original picture (the standard) and 80 subsampled images of the same picture (test images). The four standard pictures were: two natural images (Mandrill and Waves), a ray-traced computer-generated scene of moderate complexity (Gallery), and a collection of seven 1-pixel-wide lines of white and maximum-saturation colors on a black background (Lines). In each session, the standard image was shown first, and the observer was allowed to study it for as long as he desired. This was followed by five repetitions of each of the 16 pre-calculated test images in a pseudo-random order. Each test image was calculated using one of the eight subsampling parameters (n=1...8), and one of the two color spaces (YUV or CIELUV). On each trial, the observer indicated whether the test image was the "same as," or "different from" his or her recollection of the standard image, which was not displayed again in the session.

Results and Discussion

Figure 2 shows the percentage of "same" responses, pooled over all eleven observers, as a function of the chromatic subsampling parameter, n. For n=1, the test images are mathematically identical to the standard pictures, and the data are correspondingly near 100% "same" judgements for those conditions. As n increases to 8, most pictures (except Waves, see below) using both color spaces have evident color artifacts, and the data fall to near 0%. For subsampling parameters in between these extremes, whether the test image is judged to match the standard depends on which picture it is, and which color space was used. The error functions fit to the data allow us to estimate the subsampling value that results in 50% "same" judgements for each picture and color space. In the YUV space, this threshold value of n is 4 for the Gallery, 3.5 for the Mandrill, 1.5 for the Lines, and 7 for the Waves. The threshold values are 2.5, 2.5, 1.5, and 1.5 for the Gallery, Mandrill, Lines and Waves in CIELUV space.

The curves for CIELUV subsampling are uniformly to the left of the YUV curves, which indicates that a greater degree of undetected subsampling can be accomplished by using the YUV space rather than CIELUV. The Waves data, for example, fall to 0% for n=2, i.e., any spatial averaging is detectible. This is due to the common (in this picture) occurance of brightly colored sparkles in the lowlights of the image that are not in the standard. These sparkles are caused by estimating chromatic values at a dim pixel that are not possible to generate from *any* RGB triplets having that luminance. The transformation back to RGB values generates negative (or over-maximum) RGB values, which when truncated to fall within range, create "colored confetti" on the screen.

Subsampling the Lines picture was detectible for all subsampling values in both color spaces. The 2-D spatial averaging causes the bright, saturated lines of the original to desaturate, with the colors bleeding out to neighboring pixels. There is one region on the Mandrill's nose that became "blotchier" as the subsampling size increased, and all observers reported basing their discriminations on that feature. The standard Gallery has text, smooth-shaded walls, and brightly-colored round objects receding into the distance. When subsampled, these small intensely-colored objects become less saturated, and their edges become fuzzier; those were the features that most observers reported using in their discrimination. The Waves picture has a wave breaking and spraying in the middle of the photograph, with a defocused mountain in the background; it has a certain "grainy" appearance. As the subsampling increased, the graininess was averaged out, thereby increasing the apparent quality of the picture. Since the observers were instructed to detect differences, however, they judged these improved pictures as "different."

The main conclusion of the data in Figure 2 is that the success of the subsampling algorithm is very dependent on the contents of the image, and also on which color space is used to perform the calculations. A subsampling parameter of n=4 (as in Figure 1) corresponds to a factor of 16 savings in the amount of stored chromatic information; a 24-plane picture can therefore be stored in only 9 planes (8 planes achromatic + 8/16 planes for each chromatic dimension) by subsampling in a 4 by 4 grid. The results of this study show that subsampling by that factor can be detected only about half the time on the natural and ray-traced images we tested (Gallery, Mandrill, and Waves), if the averaging is done in the YUV color space. The subsampling process can almost always be detected, however, on simple graphics pictures such as the Lines.

Figure 2: Results of discrimination experiment, showing the percentage of observations judged "same as standard," as a function of the chromatic subsampling parameter, n. Part a: Gallery picture. Part b: Mandrill picture. Part c: Lines picture. Part d: Waves picture. Circles and solid lines are for the YUV condition, triangles and dashed lines are for the CIELUV condition.

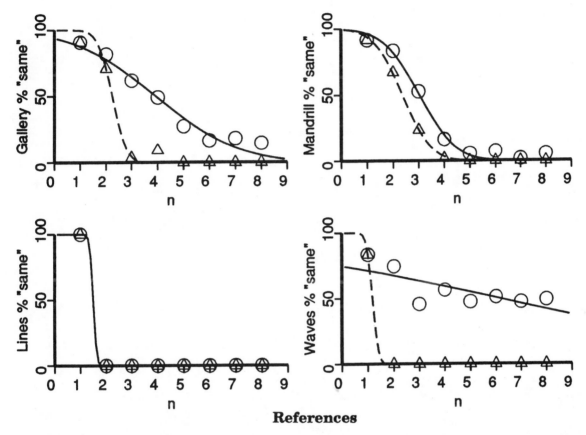

References

Benson, K. B. (ed), (1986). *Television Engineering Handbook*, New York, McGraw-Hill, 18.24-18.25.

CCIR (1983). "Encoding parameters of digital television for studios," **September 27**.

DeValois, K.T., and Switkes, E. (1983). "Simultaneous masking interactions between chromatic and luminance gratings," *Journal of the Optical Society of America*, **73**, 11-18.

Granger, E. M., and Heurtley, J. C. (1973). "Visual chromaticity-modulation transfer function," *Journal of the Optical Society of America*, **63**, 1173-1174.

Hurvich, L., and Jameson, D. (1955). "Some quantitative aspects of an opponent-colors theory. II. Brightness, saturation, and hue in normal and dichromatic vision," *Journal of the Optical Society of America*, **45**, 602.

Larimer, J., Krantz, D.H., and Cicerone, C.M. (1974). "Opponent-process additivity. I. Red/Green equilibria," *Vision Research*, **14**, 1127-1140.

Watson, A. B. (1987). "Efficiency of a model human image code," *Journal of the Optical Society of America A*, **4**, 2401-2417.

Wyszecki, G., and Stiles, W.S. (1982). *Color Science, 2nd Edition*, Wiley-Interscience.

FRIDAY, JULY 14, 1989

1:00 PM–2:30 PM

FB1–FB5

COLOR CODING

Gerald Murch, Tektronix, *Presider*

Segregation of Basic Colors in an Information Display

Robert M. Boynton and Harvey S. Smallman
Department of Psychology C-009
University of California at San Diego
La Jolla, CA 92093

Introduction. Consider a display containing many elements, among which there is a single critical target, differing slightly from the others, for which a subject is instructed to search. If a very small spatial detail defines this difference, (a) fixation near the critical target will be required for its identification, (b) a serial search will be needed to find it, and (c) the larger the number of targets, the longer will be the average search time needed to discover the critical target. The search time can be reduced if: (a) targets are color coded, (b) some of the targets are of an irrelevant color, and (c) the subject knows the color of the critical target. As Green and Anderson (1956) first put it, "When Os know the color of the target, search time is approximately proportional to the number of symbols of the target's color." Their seminal research, involving only two colors, has subsequently been extended to five by the work of Smith (1962).

However, in none of the studies to date, has the rationale for choice of colors been made explicit, or have more than five colors been employed. In 1969, the anthropologists Berlin and Kay introduced the concept of basic color terms. Subsequent research with surface colors, some of it in this laboratory, has indicated that there are eleven basic color sensations, each identified with a well-learned name. The use of these terms differs from that of all other color terms (termed nonbasic) in several ways including greater speed, accuracy, and consensus in naming colors with them (Boynton and Olson, 1987). Based on color naming data of 27 subjects, focal examples of basic colors were selected from among the 424 colors of the OSA Uniform Color Scales set (Nickerson, 1981), and members of an independent group of 24 subjects were asked to name them using only basic color terms; perfect consensus resulted. The OSA specification of these color samples, the names used, and the number of subjects in each group who used that name, are given in Table 1.

The present study seeks to determine the extent to which focal basic colors segregate in an information display.

Experimental trials. On most of more than four thousand trials of the experiment for each subject, a critical target was present, whose color was one of 10 focal basic colors (all except gray) seen against a gray background, located among an unpredictable number of irrelevant (decoy) targets. Decoys were squares subtending 1 deg of visual angle at the viewing distance of 57 cm. To create the critical target, a 3 x 4.5 min notch was removed from a corner of one of the squares. The subject responded when the critical target was discovered and his response time was measured. (The subject was not required to indicate either the location of the critical target in the array, or which corner was notched.)

With the subject looking at a gray screen subtending 28 by
36 deg of visual angle, a trial begin with a 1-sec presentation
of information concerning the color of the critical target to be
presented, provided either as a color name (e.g., RED) in black
capital letters) or as one of the focal colors presented as as a
1-deg square. The critical target then appeared in an
unpredictable location, with an unpredictable corner notched,
somewhere among an array of decoy targets. Immediately upon the
subject's response (or after 12 sec if there was none by that
time) the display returned to gray for about 3.5 sec until the
next trial.

On 400 trials, no knowledge of the critical target color was
given. Instead, the letter X was seen where the color
information usually appeared. Additionally, on any given trial
there was a 10 pct chance that a critical target would be absent.
If the subject incorrectly responded, an auditory signal was
provided.

Experimental design. Each display consisted of 10 targets
having the color of the critical target, and from 10 to 90
targets of different colors, also in groups of 10. Not all
possible combinations of critical and irrelevant target colors
were possible. We therefore decided to choose the irrelevant
targets at random. For example, when red was the target color,
a series of 10 conditions was prepared that might look like this:

```
RED
RED,GREEN
RED,GREEN,PINK
RED,GREEN,PINK,YELLOW
RED,GREEN,PINK,YELLOW,BROWN
RED,GREEN,PINK,YELLOW,BROWN,BLUE
RED,GREEN,PINK,YELLOW,BROWN,BLUE,ORANGE
RED,GREEN,PINK,YELLOW,BROWN,BLUE,ORANGE,PURPLE
RED,GREEN,PINK,YELLOW,BROWN,BLUE,ORANGE,PURPLE,BLACK
RED,GREEN,PINK,YELLOW,BROWN,BLUE,ORANGE,PURPLE,BLACK,WHITE
```

Then another such set was prepared, starting with RED and adding
the other colors in a different random order. Altogether, 20
such sets were be prepared for RED. Then 20 sets of this kind
were prepared for GREEN, and then also for the other 8 basic
colors. Thus there was a total of 2,000 such sets, 200 for each
search color, 20 for each number of colors among which the search
color might be displayed. A given experimental session consisted
of 200 normal trials, including 20 critical-target color trials
for each color, with from 0 to 90 decoys presented twice, once
for each type of color information. In addition, there were 10
no-information trials and up to 21 (10 pct) no-target trials.
The trials occurred in a different random order for each session.
Twenty sessions, each of about 40 min duration, were required to
complete the design for each of two subjects.

Technical details. Stimuli were presented on a computer-
controlled monitor (Macintosh II, TAXON Ultra Vision 1000) with
just over 1 million pixels in its 28 x 36 cm display. Target
colors were determined by temporarily setting up a divided field,
half of which was filled with an OSA color sample,
spot-illuminated by a Carousel projector through the a blue

filter taken from a Macbeth easel lamp.   In the other half of the field, a color produced on the monitor was seen which could be continuously varied by an experimenter, to achieve a match. This was done for each of the 10 focal OSA colors except gray, and also for a sample of the flat 20-pct gray background color used in our previous color-naming studies.   Each of two experimenters made 5 matches for each color, and the R,G,B gun-values were averaged for use in the main experiment.

At the viewing distance chosen, 1 cm on the display corresponded exactly to 1 deg of visual angle, with 40 pixels per deg.   Targets were restricted to a 16 x 20 cm central region of the monitor,leaving a gray surround of from 6 to 8 cm.   Within the 320 possible 1-deg squares of the target area, stimuli were located randomly with the restriction that adjacent squares could not be filled with targets; touching corners were allowed.   The luminance of the gray screen area was 1 cd/m2; those of the target colors are given in Table 1.   The corner notch of the critical target was created by substituting a 2x3 pixel array of the background gray.

Results.   Figure 1 shows the results for HSS.   Mean response time is plotted as a function of the number of groups of 10 colors to be searched.   The upper curve, which represents the results for the minority of trials in which the target color was not specified, shows response time increasing from about 2 sec to more than 10 sec.   Because of the limit of 12 sec of search time, the curve would have risen even higher had unlimited search time been permitted.

The lower curves are for the majority of trials on which information about target color was provided.   The virtual overlap of the two sets of data shows that it makes no difference whether the color information is provided by word or by sample.   Response times are all short, and increase by only about 1 sec over the range from 1 to 10 color groups.

A second subject (RMB), whose data are not shown here, shows very similar results.   His response times are about a second longer for the 1 color group condition, and they rise by about a second and a half as the number of color groups is increased to 10.   His curve for the no-information condition is also similar to that for HSS.

Discussion.   Our results confirm that, in a search task where the color of a critical target is known, targets of irrelevant color can largely be ignored even when there are as many as 9 different kinds of irrelevant colors.   The colors used were chosen to match focal examples of basic colors.   In other words, basic colors segregate very well in an information display.   Not revealed in the data plotted here, however, are the following complications:

● Response times for the different colors varied substantially, and the rate of increase of response time with the number of added sets of irrelevant colors was not the same for all colors used.

● Not all colors appeared to segregate equally, although we have no measure of this.

● The detectability of the notch on the critical target was a function of target color. In particular, green critical targets were difficult to discern.

These complications will be discussed in more detail at the meeting, and it is expected that additional experimental data related to them will be available by then.

References

Berlin, B. and Kay, P. (1969) Basic Color Terms. Berkeley: University of California Press.

Boynton, R.M. and Olson, C.X. (1957). Locating basic colors in the OSA space. Color Research and Application 12, 94-105.

Green, B.F. and Anderson, L.K. (1956). Color coding in a visual search task. J. Experimental Psychology 51, 19-24.

Nickerson, D. (1981). OSA Color Scale samples: a unique set. Color Research and Application 6, 7-33.

Smith, S.L. (1962). Color coding and visual search. J. Experimental Psychology 64, 434-440.

Table 1. Specification of the basic colors.

Color Name	Luminance (cd/m2)	OSA Coordinates L j_ g			Number of subjects using name Exp 1 (N=27)	Exp 2 (N=24)
red	0.59	-4	2	-8	23	24
green	1.63	-3	5	5	25	24
yellow	5.58	4	12	0	25	24
blue	0.59	-4	-4	4	23	24
orange	1.85	1	9	-7	25	24
purple	0.28	-6	-4	-2	23	24
pink	3.49	3	-1	-5	23	24
brown	0.54	-5	3	-3	25	24
white	8.27	---			--	24
black	0.01	---			--	24

Notes:

All other color names in Exp. 1 were nonbasic terms.

White and black were not selected from Exp. 1. Instead, for Exp. 2 were they were provided by a piece of white paper and carbon paper respectively. A somewhat darker green than the one specified here was used.

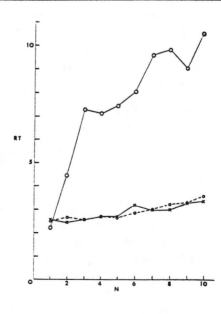

Figure 1. Response time in seconds as a function of the number of groups of 10 targets in the display. All targets within each group were of the same color, and colors differed between groups. Large open circles and solid line: trials on which information about the color of the critical target was not given. Color information was provided for the other conditions. Small circles, dashed line: Examples of the critical target color were given. X's, solid line: Color words were displayed instead of examples.

Color and Visual Search in Large and Small Display Fields

Allen L. Nagy, Robert R. Sanchez, and Thomas C. Hughes
Psychology Department, Wright State University
Dayton, Oh. 45435

In previous experiments we have attempted to measure color differences required to obtain parallel visual search. (Nagy and Sanchez, 1988) The observer's task was to search for a small target disk that differed only in color from white distractor disks also present in the display. When the color difference between the target and distractors was small the time required to find the target depended on the number of distractors present. We refer to this as a serial search. When the color difference was large the search time was independent of the number of distractors present. We refer to this as parallel search. In order to estimate the color difference required to obtain parallel search we measured search time as a function of color difference in displays with a large number of distractors. Search times decreased with increases in color difference until a minimum was reached and then remained constant with further increases in color difference. From these experiments we estimated the smallest color difference at which parallel search could be obtained along several lines in color space. We will refer to these color differences as critical color differences, or CCD's. CCD's varied from approximately 12 to 40 times the size of the MacAdam unit of color difference depending on the direction in color space.

In the experiments to be described here, we explore a possible explanation (Triesman and Gelade, 1980) for the serial and parallel searches obtained with small and large color differences. In previous experiments, the target and distractors were small disks, 1/2 degree in diameter, presented in random positions in a dark field 15 degrees wide and 11 degrees high. Though the observer was free to make eye movements, a large proportion of the stimuli fell outside the fovea on peripheral retina at any given point in time. It is well documented that the ability to discriminate color is poorer for stimuli presented outside the fovea and that the discrimination threshold continues to increase as stimuli are moved further and further into the periphery. When the color difference between target and distractors was small in our earlier search experiments, the difference may have been below threshold for peripheral portions of the retina. The observer would then have to fixate the target foveally in order to determine whether it was different than the distractors. Since the target placement was random, the observer would be required to fixate random portions of the display or perhaps

in chromaticity in the Macleod and Boynton (1979) chromaticity
diagram. Previous work had shown that mean log search time
decreased approximately linearly as a function of the
difference between target and distractors up to a point and
then remained roughly constant with further increases in the
difference. Therefore we chose four difference levels so that
they fell on the decreasing portion of the curve and the fifth
level was chosen to be the largest difference obtainable along
the chosen line. Straight lines were fit to the decreasing
search times and a horizontal line was drawn through the mean
minimum search time. The intersection of the two lines gives
an estimate of the CCD. Results for the small and large
displays were quite similar as can be seen in the figure.
CCD's for five of the six directions in color space differed
by less then 12% for these two conditions. In the sixth
condition the target chromaticity fell on a tritan confusion
line through the distractor chromaticity and the targets
appeared to be yellower in hue. The CCD for the small display
field was 50% larger than the CCD for the large display in
this condition. We are reexamining this condition. When the
small disks were used with the large display field, CCD's were
much larger for all six directions in color space. They were
1.5 to 2.0 times larger than CCD's in the other two conditions
depending on the direction in color space.
     The results suggest that restricting the display field to
an area approximately the size of the fovea does not
necessarily reduce the size of the CCD. This result would
seem to argue against the peripheral loss hypothesis, which
would predict that CCD's should decrease in size as the
display area is reduced. However there is a complicating
factor. In order to present the same number of stimulus disks
in a smaller display area it was necessary to reduce the size
of the stimulus disks as well. We arbitrarily reduced both the
dimensions of the display field and the diameter of the
stimulus disks by a factor of four. This may have been a
fortuitous choice. Some other choice of reduction factors
probably would not have resulted in similar CCD's for the two
conditions. Comparison of the results obtained with the small
display field with the results obtained with small disks in a
large field suggests that reducing the display area does
reduce the CCD, as predicted by the peripheral loss
hypothesis, if stimulus size is held constant. Both the size
of the display area and the size of the stimuli appear to be
important factors in determining the size of the CCD. Reducing
the size of the display field reduces the size of the CCD if
stimulus size is held constant. However, reducing the size of
the stimuli increases the size of the CCD. When both field
size and stimulus size are reduced, the two factors appear to
work against each other and CCD may not change in size. We are
planning further experiments to examine the peripheral loss
hypothesis and the mechanisms underlying the transition from
serial to parallel search with increasing color differences.

adopt some systematic fixation pattern until the target disk fell on the fovea. For larger color differences the target and distractors were presumably discriminable over larger portions of the retina. Mean search time should therefore decrease with increasing color difference, because the probability of the target falling within the discriminable area on any single fixation would increase. Once the color difference is large enough so that the target and distractors are discriminable over the entire retinal area subtended by the display, parallel searches should occur and search time should reach a minimum. In this view serial searches occur simply as a result of the need for foveal fixation and parallel searches occur whenever the color difference is large enough to be discriminable over the entire retinal area subtended by the display. This hypothesis predicts that as the retinal area subtended by the display becomes smaller the color difference required to achieve parallel search should become smaller, since the stimuli are restricted to a retinal area in which threshold is lower. We have attempted to test this hypothesis, which will be referred to as the peripheral loss hypothesis, by comparing the size of the critical color difference in small and large displays.

The large display was 15 degrees wide and 10 degrees wide with stimulus disks 1/2 degree in diameter. The small display was circular with a diameter of 4 degrees. The stimulus disks were also reduced in size to a diameter of 1/8 degree so that the same number of stimulus disks (54) could be presented in the small display field as in the large display field. The reduction in the size of the stimulus disks introduces an additional complication, since it is well documented that color discrimination thresholds increase with decreases in the size of the stimulus field. Therefore we included a third condition in which the small disks (1/8 degree diameter) were presented in the large display field. Critical color differences were measured in both directions along tritan and a deutan confusion lines through the distractor chromaticity (x=.373, y=.383), and also for increases and decreases in luminance. The distractor color was approximately white. Four observers with normal color vision took part in the experiments. All had some practice at the task prior to data collection.

Typical results for two of the six directions in color space are shown in Figure 1. Log search time in milliseconds is plotted as a function of the difference between the target and distractors. Mean results from four observers are shown for the three display conditions in each panel. In the panel on the left, the target differed from the distractors in that its luminance was higher. Values on the abcissa indicate the difference in log luminance. In the panel on the right the target chromaticity fell on a deutan confusion line through the distractor chromaticity and targets appeared to be redder than the distractors. The abcissa represents the difference

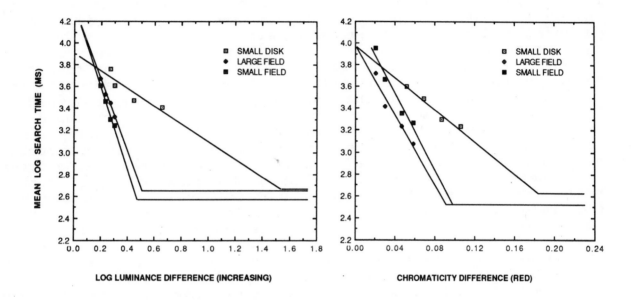

Figure 1. Mean log search times as a function of color
         difference.

References

MacLeod, D.I.A., and Boynton, R.M. (1979) Chromaticity diagram
     showing cone-excitation by stimuli of equal luminance.
     J. Opt. Soc. Am., 69, 1183-1186.

Nagy, A.L., and Sanchez, R.R., (1988) Large color differences
     measured with a visual search task. Invest. Ophthal. and
     Visual Sci. (Supplement) 29, 407.

Nagy, A.L., and Sanchez, R.R., (1988) Color differences
     required for parallel visual search. Optics News, 14,
     137, Abstract.

Triesman, A., and Gelade, G., (1980) A feature integration
     theory of attention. Cognitive Psych., 12, 97-136.

<u>Theoretical Constraints on the Participation of Rods and Cones in Color Matches</u>, Michael H. Brill, Science Applications Interational Corporation, 803 West Broad St., Suite 100, Falls Church, Virginia 22046.

1. <u>Introduction</u>. The visual system is trichromatic for small, centrally fixated fields and high (photopic) light intensities. For such conditions, color matching requires three primary lights, because of the three kinds of cone photoreceptors involved. When light intensities are low (scotopic), color matching reduces to brightness matching, and vision becomes monochromatic. At scotopic light intensities, which are not enough to excite cones, vision is mediated by rods; on the other hand, photopic light intensities are high enough to saturate the rods so that only cones contribute to color matches. But what of light intensities (mesopic) that are above the cone threshold but below the rod-saturation intensity? If rods and cones participate together in mesopic vision, shouldn't mesopic color matches require four primary lights instead of three? To address this question, this paper first shows the possibility of a full tetrachromatic matching space by demonstrating that the rod spectral-sensitivity function is linearly independent of a set of color-matching functions for the cones. Secondly, the mathematical conditions for convergence of Trezona's [1] iterative tetrachromatic matching experiment are derived and shown to agree with the actual conditions of the experiment. This step is intended to show that some experimental results are consistent with the formal discussion. Finally, a simple diagram is introduced that represents confusion loci in a reduced "matching space" of rod versus cone stimulation. The topology of this diagram is used to show how mesopic trichromacy is possible.

2. <u>The spectral independence of rods and cones</u>. To show the spectral independence of rods and cones, a computation was performed to find the least-square best fit of linear combinations of Judd's 1951 color-matching functions to the CIE 1951 $V'(\lambda)$ function representing scotopic sensitivity [2]. The results, displayed in Figure 1, show that the best fit of the Judd color-matching functions (solid curve) is still quite far from the $V'(\lambda)$ function (dashed curve). Hence the rod spectral sensitivity is linearly independent of the cone spectral sensitivities (represented by Smith and Pokorny [3] as linear combinations of Judd's 1951 color-matching functions).

3. <u>Conditions for convergence of Trezona's iterative matching</u>. P. Trezona [1] introduced a technique of iterative color matching that uses four primary lights to ensure a match with a test light under both photopic and scotopic conditions. First, a subject matches a photopic test light $S(\lambda)$ by adjusting the intensities of three photopic primary lights $p_i(\lambda)$ (i=1,2,3). After this photopic match is made, all the light levels are reduced to scotopic, and the subject dark-adapts and then adjusts the intensity of a scotopic primary light $p_0(\lambda)$ until a scotopic match is achieved. All lights are scaled up to photopic levels,

and the photopic primaries are then adjusted for a photopic match. Then the intensities are scaled again to scotopic, and the scotopic primary is adjusted. This procedure is repeated until no significant adjustments need to be made to preserve the match either at the scotopic or photopic levels.

The mathematical convergence of this algorithm can be addressed as follows. At iteration n, denote as three-vector $\underline{x}(n)$ the intensities of the photopic primary lights needed to ensure a photopic match; also, denote as $y(n)$ the intensity of the scotopic primary light needed to restore a match at the scotopic level at step n. Then the respective photopic and scotopic matching equations at step n are

$$\underline{a} = A\,\underline{x}(n) + y(n-1)\underline{B} \; ; \quad b = \underline{c}^T\underline{x}(n) + dy(n) \; . \qquad (1)$$

Here, $\underline{a}$ is the vector $\langle S\underline{q}\rangle$ of tristimulus values of test light spectral power distribution S ($\underline{q}$ is the 3-vector of photopic color-matching functions, and $\langle\ \rangle$ denotes wavelength integration); $A = \langle\underline{q}\,\underline{p}^T\rangle$ is the 3x3 matrix of tristimulus values of the photopic primaries; $\underline{B} = \langle p_0\,\underline{q}\rangle$ is the vector of photopic tristimulus values of the scotopic primary light; $b = \langle S\,q_0\rangle$ is the scotopic unistimulus value of test light S; $\underline{c} = \langle q_0\,\underline{p}\rangle$ is the vector of scotopic unistimulus values of the photopic primary lights; $d = \langle p_0\,q_0\rangle$ is the scotopic unistimulus value of the scotopic primary; $y(0) = 0$ as an initial condition for the experiment; and all vectors are column 3-vectors unless explicitly transposed.

Equation (1) can be rewritten as a difference equation for $y(n)$:

$$y(n) = f + g\,y(n-1) \; , \qquad (2)$$

where $fd = b - \underline{c}^T A^{-1}\,\underline{a}$, and $gd = \underline{c}^T A^{-1}\,\underline{B}$. The fixed point of this difference equation is $y = f/(1-g)$. Does $y(n)$ evolve toward or away from this fixed point? Convergence (evolution toward the fixed point) is ensured if and only if $|g| < 1$. Because $p_0(\lambda)$ and $q_0(\lambda)$ are nonnegative, the condition $|g| < 1$ means that

$$\langle q_0\,\underline{p}^T\rangle\,A^{-1}\,\langle\underline{q}\,p_0\rangle \; < \; \langle q_0\,p_0\rangle. \qquad (3)$$

The left-hand-side of this equation is the rod response to that combination of photopic primaries that is metameric (via the cone system) to the rod primary p_0. Hence convergence requires that the rod primary must be brighter to the rods than the cone-matching linear combination of the cone primaries is to the rods. This condition is ensured if the scotopic primary is near the peak of the rod sensitivity in wavelength, and the photopic primaries should be nearer the peaks of the cone sensitivities than to that of the rod sensitivity. This situation is exactly true for the primaries chosen by Trezona: The scotopic primary was at 509 nm, near the peak of the rod sensitivity, and the photopic primaries were at 468, 588, and 644 nm, all wavelengths of much lower rod sensitivity.

4. <u>The possibility of trichromacy in a four-receptor system</u>. If color matching requires no more than three primary lights, even under mesopic conditions, then there are confusion loci throughout the tetrastimulus space spanned by rods and cones. In the scotopic region, these loci are three-dimensional hyperplanes spanning the cone subspace. In the photopic region, these loci are one-dimensional lines spanning the rod subspace. In the mesopic region, some other confusion loci prevail, but these have never been specified. Visualizing these confusion loci should be facilitated by a stylized two-dimensional representation of these loci in a plot of rod stimulation versus cone stimulation. (Rod stimulation is computed as <S V'>, and cone stimulation would be computed as <SV> if the selected cone subspace were to be in the photopic luminosity direction.) Such a diagram is presented in Figure 2. The important features in the diagram are the projective and topological ones (e.g., the straight-line parts of the confusion loci, and the convexity of the confusion loci with respect to the origin of the diagram). Three important thresholds exist in the diagram: (1) T_1 is the absolute threshold for rod response, and is the bottom horizontal dashed line; (2) T_2 is the rod stimulation level at which the rod response saturates, and is the top horizontal dashed line; and (3) T_3 is the absolute threshold for cone response, and is the vertical dashed line. A vertical line for cone saturation is not represented in the diagram. (The juxtaposition of lines T_2 and T_3 was determined in a conversation with Dr. Bruce Drum of Johns Hopkins University.)

In Figure 2, the two solid regions represent <u>areas</u> of confusion. In the lower solid region, the stimulus is subthreshold to both rods and cones; any point in this region matches absolute darkness. The upper solid region is above rod saturation but below cone threshold. The region labelled S is the scotopic, and contains confusion loci (solid lines) that are horizontal. The region labelled P is the photopic, and contains vertical confusion loci. In the region labelled ZP (zero photochromatic interval), the confusion loci are still vertical, and represent lights of wavelength long enough so that cones are activated but the rods are not. The region labelled M is the mesopic region. Visual response in the mesopic is presumably a combination of cone and rod stimulations such that increase in either stimulation moves the response to a confusion locus farther from the origin. Confusion loci in this region are not necessarily straight lines, but are organized such that a single ray from the origin (representing a single primary light) will cross all the confusion loci exactly once. This means that, in our diagram, some intensity of any primary light will match any other light. By extension, in the tetrachromatic space, three primary lights will be enough to encounter any confusion locus. Thus the possibility is shown of having trichromacy even in a tetrachromatic space such as is observed in mesopic vision.

A two-dimensional diagram to represent a four-dimensional space is also used in spacetime physics. Just as spacetime has three similar space coordinates and a dissimilar time coordinate, tetrachromatic space has three similar photopic coordinates and a dissimilar scotopic coordinate. Both in the Minkowski spacetime diagram and in our rod-cone diagram, two of the three similar coordinates are suppressed for ease of visualization.

5. Conclusions. Rods and cones are spectrally independent, and hence span a tetrachromatic matching space. The results of Trezona's matching experiment are consistent with a tetrachromatic formalism. However, the rod-cone diagram shows that any color match could still be possible using only three primary lights even in mesopic conditions.

6. References. [1] P. W. Trezona, Vision Research 13 (1973), 9-25. [2] G. Wyszecki and W. S. Stiles, Color Science, 2nd Ed. Wiley, 1982. [3] V. C. Smith and J. Pokorny, Vision Research 15 (1975), 161-171.

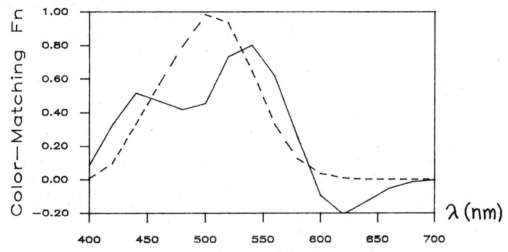

Fig. 1. V'(λ) fit by Judd's 1951 color-matching functions.

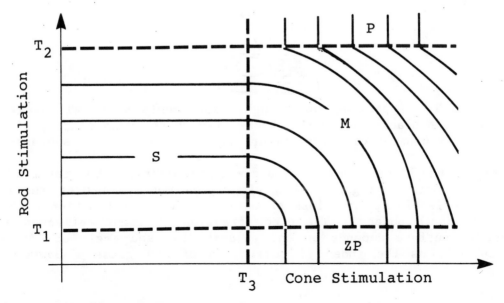

Fig. 2. Rod-cone diagram.

Light Source Size and the Stimulus to Vision

James A. Worthey, Lighting Group, National Institute of Standards and Technology, Building 226, Room A313, Gaithersburg MD 20899.

INTRODUCTION. Non-luminous objects present a stimulus to vision that depends on their optical interaction with the sources that illuminate them. "Sources" in this context means all other objects, as may be seen if the main object of interest is a plane mirror, or a mirrored sphere, for example. Such examples show that contrast in an object can depend on contrasts elsewhere in the environment. An important and highly variable feature of the optical environment is the size of the primary light source. This paper will look at light source sizes and their effect on the appearance of shiny objects. (Obviously, source size affects the appearance of matte objects also.) **Table 1** shows that familiar lights vary in the bright solid angle that they present by a factor of 10^5 or 10^6.

Table 1, Light Source Sizes.

Light Source	Area, meter^2	Solid Angle at 2 meters distance, microsteradians
Unfrosted 60 W inc. bulb	2.0×10^{-5}	5
The Sun (distance = 93×10^6 miles)	1.5×10^{18}	67
Ordinary frosted 60 W bulb	3.1×10^{-4}	79
Soft White 60 W bulb	2.4×10^{-3}	590
F40T12 fluorescent tube	4.6×10^{-2}	12,000
Luminous ceiling, extending to ∞	many	$2\pi \times 10^6 \approx 6,300,000$

VEILING REFLECTIONS. Much can be learned about the effects of light source size from *Gedanken* experiments---calculations---based on the apparatus of **Figure 1.** The white surface reflects all the light that strikes it, but

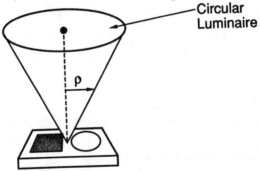

Figure 1. Apparatus for thought experiments with veiling reflections. A black glass and a Lambertian white surface are positioned directly under a circular uniform light source of semi-subtense ρ.

diffusely; the black glass gives a clear mirror-image, but returns only about 4% of the incident light. This simplified model captures the essence of "veiling reflections," the loss of contrast due to reflection at the surface of a dielectric object. The exact fraction of light reflected at a shiny dielectric surface depends on the polarization and angle of incidence of a ray, as well as the index of refraction of the glass. However, for unpolarized light incident within 30°-40° of normal, it is a good approximation[1] just to say that the reflected fraction is 0.04.

Now assume that observations are made from a point near the axis in **Figure 1**, neglecting the possibility of the photometer casting a shadow or seeing its own reflection. Expressing a luminance as a fraction of white gives a "gray level." The gray level g of the light source image in the black glass is

$$g = 0.04 \ / \ \sin^2\rho \qquad\qquad (1)$$

For a light source that covers the hemisphere above the test object, $\rho = 90°$. The gray level is then 4%, which is not negligible because it may be high relative to contrasts that are being veiled, and because of the non-linearity in the way they eye sees lightness. Four percent gray level corresponds to Munsell Value 2.3.

VEILING REFLECTIONS AND COLOR. A look at veiling reflections in colored objects reinforces the view that such reflections are never negligible. The effect of veiling reflections on the gamut of colors available under Illuminant C was calculated, starting with Pointer's "gamut of real surface colors."[2] Working in the cylindrical polar form of CIELAB space, Pointer reported a locus of the most saturated colors (highest radial coordinate $c*$) with $L*$ and hue angle ($h*$) as independent variables, based on measured data. The veiling reflectance calculation found the new, desaturated value ($L*$, $h*$, $c*$) that each color would have with a given admixture of white light (due to a veiling reflection). Graphical results, which have been published[2], show that a few whitish colors are gained, but many deep colors are lost, so that there is a net reduction of the gamut volume in CIELAB space. Table 2 shows the relative volume for selected veiling reflection levels.

Table 2, Volume of Color Solid as a Function of Veiling Luminance

Veiling Reflection, As % of White	Relative Volume of Color Solid, as %
0 %	100 %
4	63
8	46
12	35
16	28

A "mere" 4% veiling reflection shrinks the color gamut by 37%.

SOURCE LUMINANCE. The lights in Table 1 were chosen on the basis that each approximates a familiar lighting situation. This shows intuitively how much lights can vary in size. A basic reason for this variation is that they vary tremendously in luminance. Consider the source of **Figure 1**. Let E be the illuminance at the object under the circular source, and let L be the luminance of the source. Starting with a standard formula, it is found that

$$\rho = \sin^{-1}\left\{\left[\frac{E}{\pi L}\right]^{1/2}\right\} \qquad \approx \quad \frac{180}{\pi}\left[\frac{E}{\pi L}\right]^{1/2} \qquad\qquad (2)$$

$$\Omega = 2\pi(1 - \cos\rho) \quad \approx \frac{E}{L} \quad . \tag{3}$$

Here Ω is the solid angle subtended by the source. In Equations (2) and (3), the expression directly to the right of "=" is exact, while the expression after "≈" depends on the small-angle approximation.

Eq. (2) and (3) are graphed in **Figure 2**, assuming E = 1000 lux. Points on the graph of solid angle represent some familiar light sources, based on luminances given by Wyszecki and Stiles[3].

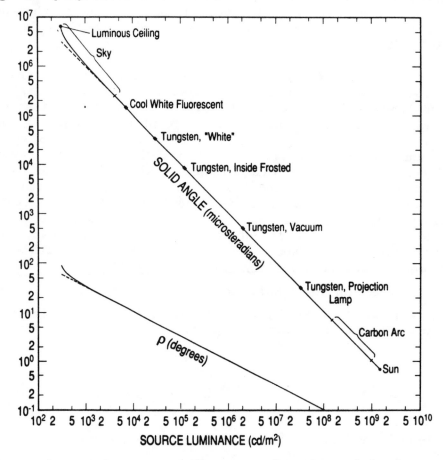

Figure 2. Light source size as a function of luminance, for E = 1000 lux.

HIGHLIGHTS. Highlights are images of a light source, seen in the surface of a shiny object, whether the object is metallic or dielectric. The term is apt because even in a dielectric object, highlight luminance is high when the source is small. For instance, applying Eq. (1) when the sun is the source shows that highlight luminance is some 2000 times the luminance of a white.

Highlights are of particular visual interest in curved surfaces, even highly curved surfaces. For instance, highlights will tend to pile up on the rim of a glass. If the source is small, this highlight region gives high contrast and serves to localize sharply an important detail of the glass. As for the luminance of such a highlight, it is equal to the source luminance times a reflected fraction that can be taken as 0.04 for dielectric objects[3] (as in Equation 1). Highlight luminance does not depend on surface curvature, so long as the surface is shiny. However, luminance loses its visual significance when the image falls below the limit of visual resolution. Table 3 shows how, as the source subtense increases, the object radius must decrease for the highlight to appear as a point. The object is 0.5 m from the eye and 2 m from the light. The message is that the sun will appear as a point in many but not all common objects, while bigger sources will give fat highlights, except in tiny objects.

Table 3: Object radius for highlight to be seen as a point.

Light Source (2 meters distant)	Subtense, min arc	Object radius for $\alpha = 1$ min arc, mm
The Sun (dist = 93×10^6 miles)	32	31
Unfrosted 60 W inc. bulb	34	29
Soft White 60 W inc. bulb	95	11
F40T12 fluorescent tube	2034	0.5
Luminous ceiling, extending to ∞	10800	no highlights

DYNAMIC RANGE. The luminous ceiling gives nothing that can be called a highlight, only a veiling reflection of gray level 0.04. With a small source, highlights add to the contrast and dynamic range of a scene. Consider the setup of Figure 1, but assume realistically that there is a slight haze on the surface of the black glass, so that its gray level away from the source image is 0.01, rather than 0.0. This leads to Table 4:

Table 4: Highlight Luminance & Dynamic Range

Light Source	Hilite Lum.	Dyn. Range
Unfr. 60 W	25,000	2,500,000
Sun	1900	190,000
Ord. Fr. 60 W	1600	160,000
Soft White	210	21,000
1 Fluor. Tube	11	1100
Lum. Ceiling	.05	20

Spaces lit by a bright ceiling full of fluorescent lamps may look "washed out." **Table 4** suggests that this is not mysterious. They look washed out because they *are* washed out, with no highlights and no jet blacks or deep colors. Loss of color contrasts due to poor color rendering will compound the ill effects of the large, dim light source[4].

References

1. James A. Worthey, "Geometry and amplitude of veiling reflections," *J. of the I.E.S.* **18**(1):49-62 (1989).

2. Michael R. Pointer, "The gamut of real surface colors," *Color Res. Appl.* 5, 145-155 (1980).

3. Gunter Wyszecki and W. S. Stiles, *Color Science*, (Wiley, New York, 1967).

4. James A. Worthey, "Opponent-colors approach to color rendering," *J. Opt. Soc. Am.* 72, 74-82 (1982).

Detection of spatial-frequency selected color shifts and the contrast sensitivity functions for CRT primaries.

Hirohisa Yaguchi, Hidemi Takahashi and Yoichi Miyake
Department of Image Science and Engineering
Faculty of Engineering, Chiba University
1-33 Yayoicho, Chiba 260, Japan

1. INTRODUCTION

The contrast sensitivity function for human vision is well understood. It is known that the contrast sensitivity of luminance discrimination shows a band pass profile and that of chromatic discrimination shows a low pass profile. This characteristics of human vision is applied to the NTSC television transmission system. The image on color TV is formed by the basis of additive mixture of three primary colors; red, green and blue. When we estimate the image quality of color TV, therefore, it becomes very important to know the spatial frequency characteristics of each primary color for human vision.

In this paper, we investigate the relationship between the contrast sensitivity functions of the primary colors and the detection of spatial frequency selected color shifts on a color CRT display.

2. EXPERIMENTAL APPARATUS

The experiment was done with a computer controlled color CRT monitor (SONY PVM-1371Q). The intensity of each primary is quantized in 8 bits, that is, 256 levels. Spatial resolution of the display is 28 pixels per cm. The chromaticity coordinates of the primaries are (0.644, 0.328) for red, (0.283, 0.597) for green and (0.151, 0.053) for blue. The colorimetry of the monitor was done with the method developed by Cowan[1]. The distance from the observer to the monitor was set to 140 cm. At this seeing distance, one pixel corresponds to 51 seconds of arc.

3. CONTRAST SENSITIVITY FUNCTIONS

A horizontally sinusoidal pattern as shown in Fig. 1 was displayed on a CRT monitor. One of the primary colors, red in the case of Fig. 1, was modulated in luminance. The other two primary colors were spatially uniform. The mean chromaticity coordinate of the test pattern was set to the D65 white whose luminance is 50 cd/m^2. The size of the test pattern was 4° x 0.5°. The observer adjusted the modulation of the test primary luminance to determine the threshold of breaking the uniformity of the pattern. The test pattern was observed binocularly. The contrast C was defined as

$$C = (L_2 - L_1) / (L_2 + L_1), \qquad (1)$$

where L_2 and L_1 are the maximum and the minimum luminance of the test primary color at threshold respectively. The contrast sensitivity for a primary color was defined as the inverse of the threshold contrast.

Fig. 2 shows the mean contrast sensitivity functions for three primary colors. These curves were derived from five observers. The result shows that the contrast sensitivity for green primary is the highest and that for blue primary is the lowest. The contrast sensitivity curves consist of two curves separated at a dip frequency around 6 cpd. The observers reported that they detected chromatic difference at the low frequency region up to 4 cpd. At the high frequency region, however, they could not see any chromatic difference but detected achromatic difference.

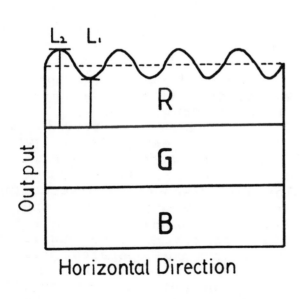

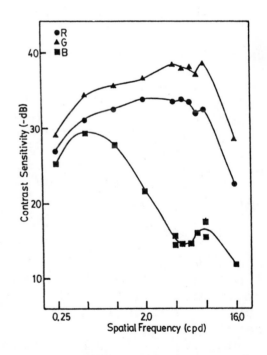

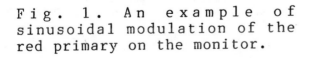

Fig. 1. An example of sinusoidal modulation of the red primary on the monitor.

Fig. 2. Contrast sensitivity curves for three primary colors of the monitor.

4. DETECTION OF SPATIAL-FREQUENCY SELECTED COLOR SHIFT

The purpose of this experiment is to investigate how the contrast sensitivity functions affect the image quality. A white random dot pattern was used for an original pattern. The power spectrum of the pattern was uniform over the spatial frequency, and the intensity histogram of the pattern was also uniform. The chromaticity of the pattern was same as the D65 white. That is, red, green and blue images were of the same pattern. A series of test patterns was prepared for subjective judgments of detection of color shifts. In the present experiment, the color shifts were made in one dimension, that is, a horizontal direction. The test patterns were made as follows. Each primary test pattern $I_i(x)$ was Fourier transformed to $A_i(u)\exp[j\theta_i(u)]$, where i corresponds to each primary color, and $A_i(u)$ and $\theta_i(u)$ are, respectively, the amplitude and the phase of the Fourier component at the spatial frequency u. Then an arbitrary selected spatial region of one of three primaries was shifted in phase as follows.

$$\theta_i'(u) = \theta_i(u) + 2\pi K/T(u), \qquad (2)$$

$$u_0/\sqrt{2} < u < \sqrt{2}u_0, \qquad (3)$$

where K is an arbitrary number of pixels which corresponds to the width of shift, and T(u) is the number of pixel in a cycle, and u_0 is an operated spatial frequency for the color shift. The value of K was set to 5 for all operated frequencies in this experiment. $A_i(u)\exp[j\theta_i'(u)]$ was inverse Fourier transformed to $I_i'(x)$ which was used for the test pattern.

The subjective estimation of color shifts were done by the paired-comparison method. Two different test patterns were positioned vertically on the monitor. The observer determined which pattern was more detectable for color shifts. The observer rating value for detection of color shifts was derived according to the Case V of the low of comparison judgments by Thurstone[2].

The estimation was divided two sets. In the first set, the operated spatial frequency was fixed to one of five spatially frequencies; 0.5, 1, 2, 4, and 8 cpd and the operated primary color was changed. The results are shown in Fig. 3. For all spatial frequencies, color shifts of green primary were the most detectable. It is clear that blue color shifts were hard to be detected. In the second set, the operated primary color was fixed and the operated spatial frequency was changed. Fig. 4 shows the observer rating values as a function of spatial frequency. For red and green primaries, color shifts at 2 cpd is the most detectable.

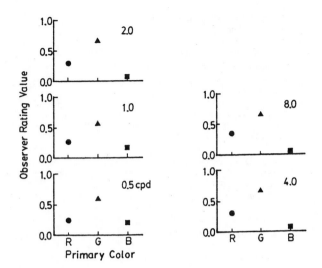

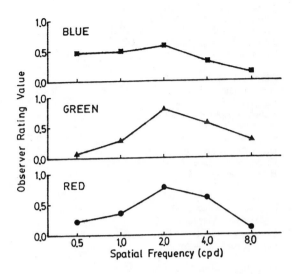

Fig. 3. Observer rating values for detectability of color shifts as a function of primary color.

Fig. 4. Observer rating values for detectability of color shifts as a function of spatial frequency.

5. DISCUSSION

The peak of the contrast sensitivity functions for red and green primaries is about 4 cpd. On the other hand, the maximum observer rating values for detection of color shifts of red and green primaries occur at 2 cpd. It seems to be that the difference between two peak frequencies is caused by the different mechanism. The contrast sensitivity at 4 cpd can be determined by the achromatic channel, on the other hand, the color shifts can be detected by the chromatic channel.

In conclusion, color shifts of the blue primary was mostly hard to be detected. This property is consistent with the poor contrast sensitivity for blue primary compared with red and green.

REFERENCES

1. W.B.Cowan, "An inexpensive scheme for calibration of a colour monitor in terms of CIE standard coordinates," Computer Graphics **17**, 315-321 (1983).
2. C.J.Bartleson, Optical radiation measurements, vol.5, Academic Press, pp.455-456 (1984).

Abruzzi, RuthAnn — FA7
Adelson, Edward H. — WB2
Ahumada, Albert J., Jr. — WC, WC3
Alter-Gartenberg, Rachel — WD3
Arditi, Aries — ThB3

Barten, P. G. J. — ThA4
Barth, Brian — FA3
Bergen, J. R. — ThC2
Boynton, Robert M. — FB1
Brainard, David H. — FA2
Brill, Michael H. — FB3
Buchsbaum, Gershon — WB3
Budge, Scott E. — ThA6
Bybell, Dan — FA5

Carlson, Curt R. — ThB, ThC2
Carney, Thom — ThA5
Cornsweet, Tom N. — WD2
Cowan, William B. — FA
Crow, Frank C. — WC2

Daly, Scott — WD5
De Ridder, Huib — ThA2
Donovan, Bill — FA5

Farrell, Joyce E. — ThB5
Fitzhugh, Andrew E. — ThB5

Girod, Bernd — ThA8
Glenn, Karen G. — ThC1
Glenn, William E. — ThC, ThC1
Goelz, U. — ThC4
Gomer, Frank E. — FA6
Gould, John D. — ThB2

Hanna, E. — WD1
Hitchnert, Lewis E. — ThA7
Huck, Friedrich O. — ThA9
Hughes, Thomas C. — FB2

Infante, Carlo — ThA

John, Sarah — ThA9

Kauff, P. — ThC4
Kelly, D. H. — WA1
Klein, Stanley — ThA5
Knoblauch, Kenneth — ThB3
Krantz, John H. — FA6

Landy, Michael S. — WD4
Larimer, James — WA
Legge, Gordon E. — ThB1
Levitan, Bennett — WB3
Lippman, Andrew — ThC5

Makous, Walter — WB
Maloney, Laurence T. — WC4

McCann, John — FA5
McCormick, Judith A. — ThA9
McGuire, Michael D. — WB4
Meyer, John — FA3
Miyake, Yoichi — FB5
Mulligan, Jeffrey B. — WC3
Munson, James — FA7
Murch, Gerald — FB

Nagy, Allen L. — FB2
Narayanswamy, Ramkumar — WD3, ThA9
Nishizawa, Taiji — ThC3

Parish, David H. — WD4
Pearson, D. E. — WD1
Pelli, D. — ThB4
Pratt, William K. — WD

Roetling, Paul G. — ThA3
Roufs, Jaques A. J. — ThA2

Sanchez, Robert R. — FB2
Schaefer, R. — ThC4
Sigel, Claude — FA7
Silverstein, Louis D. — FA6
Simoncelli, Eero P. — WB2
Smallman, Harvey — FB1
Sperling, George — WD4
Stein, Charles S. — ThA7
Stone, Maureen C. — FA1

Takahashi, Hidemi — FB5
Thornton, Jay — FA5

Ulichney, R. — FA4

van Meeteren, Aart — ThA1

Wandell, Brian A. — FA2
Watson, Andrew B. — WB1, ThA7
Westerink, Joyce — ThA2
Williams, David R. — WC1
Worthey, James A. — FB4

Yaguchi, Hirohisa — FB5
Yeh, Yei-Yu — FA6
Yellott, John I. — WD2

APPLIED VISION

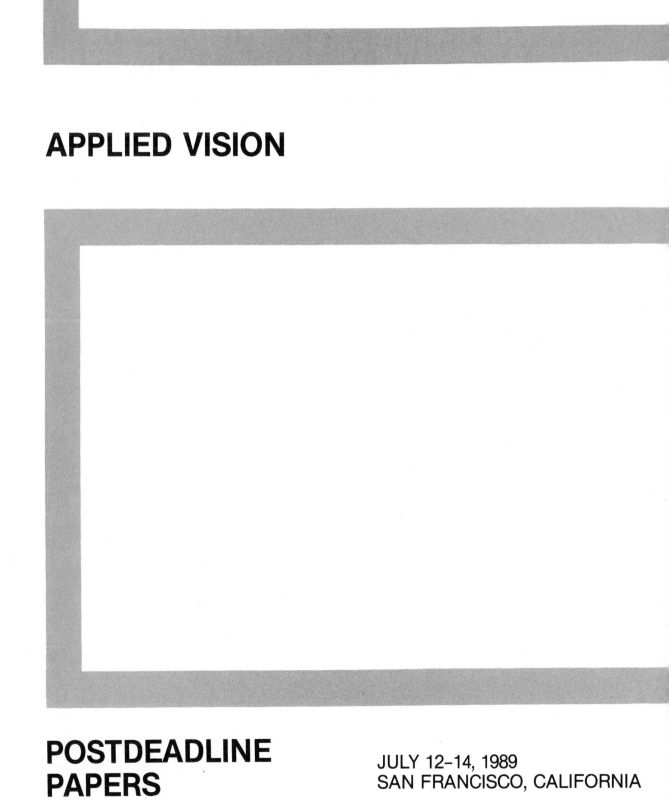

POSTDEADLINE PAPERS

JULY 12–14, 1989
SAN FRANCISCO, CALIFORNIA

CONTENTS

COLOR-NAME BOUNDARIES FOR COLOR CODING

David L. Post
Armstrong Aerospace Medical Research Laboratory
Wright-Patterson Air Force Base, OH 45433-6503

Christopher S. Calhoun
Systems Research Laboratories, Inc.
Dayton, OH 45440-3696

INTRODUCTION

One of the main problems that arises when designing color codes for electronic visual displays involves color selection. The colors must be distinctive and immediately recognizable as corresponding with the color names they represent. Otherwise, their meanings may be ambiguous, thereby defeating the code's purpose. We are approaching this problem by mapping the relationship between location on the CIE 1976 uniform chromaticity-scale (UCS) diagram and population stereotypes for color naming. This information should simplify the color selection process by helping the designer avoid, for example, specifying a "red" that actually appears orange. Thus, our project can be characterized as an attempt to improve on the Kelly (1943) color boundaries and is similar with an earlier effort by Haeusing (1976). It is also related to Boynton and Olson's (1987) work on focal colors. This paper describes our method, provides an overview of six experiments we have performed, and shows some representative results.

METHOD

Subjects. All subjects were college students having 20/20 Snellen near and far visual acuity (corrected if necessary) and normal color vision. The first three experiments used 24 subjects. The second three experiments used a different group of 12 subjects.

Apparatus. The stimuli were generated by an image processor and displayed on a color CRT monitor. A computer commanded the image processor to generate each stimulus and recorded the responses, which the subjects entered at CRT computer terminals.

Stimuli. All experiments used 210 colors that fully sample the chromatic gamut of typical CRTs (see Figure 1). The colors' luminances were adjusted to approximate the brightness of CIE standard illuminant D_{65} at a luminance of 30 cd/m^2, using the equation developed by Ware and Cowan (1983). The computer calibrated all colors each night with a spectroradiometer so their measured deviations from the desired colors were held to a tolerance of +/- 2.5% in luminance and a distance of 0.0025 on the UCS diagram. The stimulus configuration, by experiment, was:

1. Two-degree solid circle on a black background.

2. 20 arc-minute open square, black background.

3. 20 arc-minute open square, D_{65} white 30 cd/m^2 background.

4. 20 arc-minute open square, black background, 350 lux cool-white fluorescent ambient illumination.

5. 20 arc-minute open square, black background, 2050 lux daylight fluorescent ambient.

6. Two-degree solid circle, black background, 2050 lux daylight fluorescent ambient.

The solid circle represents applications involving the use of large color patches. The open square represents typical display symbology. The 350-lux ambient illumination level is typical of many office settings and produces a 14 cd/m^2 background luminance on our CRT. The 2050-lux level reduces our CRT's chromatic gamut so it is similar with an airborne CRT's under 100,000 lux ambient. It produces 85 cd/m^2 background luminance. Figures 2 and 3 show our color set and the CRT's effective gamut for the two ambient illumination conditions.

Procedure. For each experiment, the subjects performed one session per day until 12 sessions were completed. All 210 colors were presented once in each session, in a different random order. Unlimited viewing time was permitted for each color.

For each trial, the subject chose one of 12 possible color names from a list, shown on the computer terminal, and pressed a corresponding key. The 12 names were derived from a pilot study (Post and Greene, 1985) and are: blue, aqua, green, yellow, peach, orange, pink, red, purple, brown, gray, and white. This set differs slightly from the 11 color names studied by Boynton and Olson (1987). We omitted "black" because it was clearly superfluous, given our stimuli. We included "aqua" and "peach" because they are useful for color coding and it seemed likely that they would yield reliable naming.

RESULTS

Our results can only be summarized here. Readers who desire more detail are referred to Post and Greene (1985, 1986) and Post and Calhoun (1988, 1989).

Response Consistency Over Time. After each experiment, we evaluated subjects' consistency to see whether it improved substantially with practice, in which case it could be appropriate to exclude early sessions from subsequent analysis. To do this, we calculated a consistency index (C) for each Subject x Session cell. C compares a subject's responses from one session to the next and ranges from 1 (if the subject responds identically in both sessions) to 0 (if all responses differ). The values were then analyzed using ANOVA, followed (when appropriate) by post-hoc paired comparisons. We typically found evidence of small, gradual increases over time, but deletion seemed appropriate only for Session 1 from Experiment 1. Mean values of C (averaged over subjects) range from 0.66 to 0.80 across the experiments. Figure 5 shows a representative plot, taken from Experiment 1.

Color-Name Boundaries. For each color, the modal response was identified and its probability of occurrence was computed. The results from Experiment 1 are plotted in Figure 4, to illustrate. The boundaries show the transitions from one modal response to another and are only convenient approximations -- we did not estimate their exact positions. Figure 6 was derived from Figure 4 by drawing boundaries that result from setting the criterion for color-naming reliability = 0.75. (The dots representing the chromaticities in these figures are sized so they represent the 0.0025 tolerance mentioned previously.) Figures like these can be used to develop colorimetric specifications and tolerances that agree with stereotypical color naming.

Comparisons across the experiments reveal changes in the color-name boundaries that agree with well-known visual phenomena: (1) reducing stimulus size from 2 degrees to 20 arc-minutes causes shifts that are indicative of small-field tritanopia; (2) a white background tends to enhance color-naming reliability by increasing stimulus saturation; (3) a white background greatly reduces the probability of "white" responses because it becomes the subjects' white reference and makes minor deviations obvious; (4) ambient illumination has the same effect; and (5) ambient also causes boundaries to shift away from the diagram's center, indicating that greater excitation purity is needed to compensate -- but the shifts are much smaller than would be predicted from simple color-mixture calculations (e.g., Figures 2 and 3).

Other noteworthy points: (1) a white background produces some boundary shifts that suggest enhancement of the colors' blue and/or red content, apparently reflecting a change in the opponent-color channel sensitivities; (2) the modal response to the red CRT gun alone is "orange" -- to get "red", some blue must be added to increase the dominant wavelength; (3) "gray" is typically used to denote desaturated mixtures of blue and green; (4) "brown" was a modal response only in the two-degree circle, 2050-lux ambient condition; (5) the ambient did not reduce color-naming consistency; and (6) for the 20 arc-minute square, "blue" naming reliability is highest for colors having dominant wavelengths near 480 nm (i.e., psychologically unique blue), even though one might expect the most robust blues to be those having the largest blue-gun content.

DISCUSSION

The most striking thing about our color-name boundaries is that, despite the various changes with viewing conditions mentioned above, they remain remarkably stable across our experiments. These experiments have encompassed a broad range of viewing conditions encountered in actual design practice. Therefore, although we might ultimately wish to provide display designers with equations that predict color naming as a function of stimulus size, etc., it does not seem necessary to wait for the highly advanced model of color vision that would be required, nor is it necessary to perform an experiment for every unique set of viewing conditions. Instead, it appears possible to identify areas on the UCS diagram that will yield reliable color naming in many applications, based on the data we already have. The next phase of our work will involve deriving and testing one standard color set for area fills and another for symbology.

ACKNOWLEDGEMENTS

This research was supported by the Air Force Office of Scientific Research (Task 2313-V2, monitored by Dr. John Tangney). The authors wish to thank Gary L. Beckler (Systems Research Laboratories, Inc.) for developing software and otherwise assisting with the project.

REFERENCES

Boynton, R.M. and Olson, C.X. (1987). Locating basic colors in the OSA space. Color Research and Application, 12, 94-105.

Haeusing, M. (1976). Color coding of information on electronic displays. In Proceedings of the Sixth Congress of the International Ergonomics Association and Technical Program of the 20th Annual Meeting of the Human Factors Society (pp. 210-217). College Park, MD: Human Factors Society.

Kelly, K.L. (1943). Color designations for lights. Journal of the Optical Society of America, 33, 627-632.

Post, D.L. and Calhoun, C.S. (1988). Color-name boundaries for equally bright stimuli on a CRT: Phase II. In SID Digest (pp. 65-68). New York, NY: Palisades Institute.

Post, D.L. and Calhoun, C.S. (1989). Color-name boundaries for equally bright stimuli on a CRT: Phase III. In SID Digest (pp. 284-287). New York, NY: Palisades Institute.

Post, D.L. and Greene, F.A. (1985). Color naming as a function of stimulus luminance, angular subtense, and practice. In Proceedings of the Human Factors Society 29th Annual Meeting (pp. 1070-1074). Baltimore, MD: Human Factors Society.

Post, D.L. and Greene, F.A. (1986). Color-name boundaries for equally bright stimuli on a CRT: Phase I. In SID Digest (pp. 70-73). New York, NY: Palisades Institute.

Ware, C. and Cowan, W.B. (1983). Specification of heterochromatic brightness matches (Tech. Report 26055). National Research Council of Canada, Ottawa: Division of Physics.

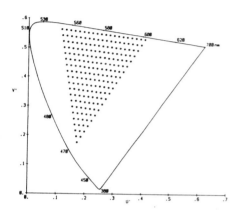

Figure 1. CIE 1976 diagram showing chromaticities used in experiment.

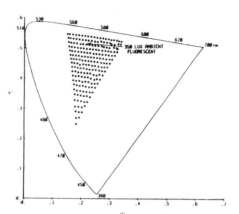

Figure 2. CRT gamut and chromaticities under 350 lux ambient.

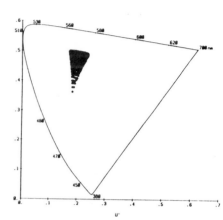

Figure 3. CRT gamut and chromaticities under 2050 lux ambient.

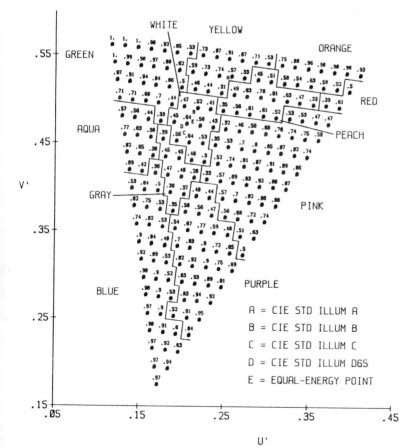

Figure 4. Exp 1 (solid circle): Probabilities of occurrence for modal responses.

A = CIE STD ILLUM A
B = CIE STD ILLUM B
C = CIE STD ILLUM C
D = CIE STD ILLUM D65
E = EQUAL-ENERGY POINT

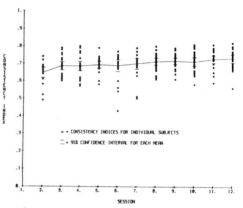

Figure 5. Consistency index for each session.

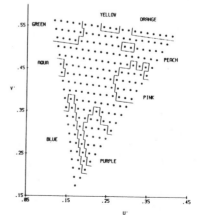

Figure 6. Color-name boundaries for $p \geq 0.75$.

PERCEPTUAL IMAGE QUALITY METRICS
(Addendum)

Jacques A.J. Roufs
Huib de Ridder
Joyce H.D.M. Westerink
Institute for Perception Research (IPO)
P.O. Box 513 - 5600 MB Eindhoven
The Netherlands

Introduction

Display engineers, specialists in image coding, image processing and related areas show a growing awareness that the quality experience of the human observer is one of the ultimate criteria of the technical quality of imaging systems. However, this awareness can only be transformed into effective action if perceptual image quality and its underlying factors, such as brightness or sharpness can be measured and related, with the physical image parameters such as luminance or bandwidth. Unfortunately the validity and efficiency of the different methods to obtain quantitative data on these perceptual attributes are still under discussion. Comparisons of different scaling techniques, comparisons of scaling results with other methods such as matchings, as well as interlaboratory tests can provide more insight into the properties of these judgement techniques. These different techniques should be applied to the same test material.
This paper concerns this kind of comparison in relation with perceptual quality and its underlying dimensions. Attention will also be paid to how the results of the measurements depend on the subjects and the selection of the scenes.
Perceptual image quality can have a different meaning in different environments (Roufs, Bouma 1980). A distinction between appreciation and performance-oriented environments is sometimes useful. HDTV images can be given as an example of the former, VDUs of the latter.
Instances stemming from both areas will be discussed.
Perceptual image quality is not easy to define (Roufs, Bouma 1980). The word implies a degree of excellence. In an appreciation-oriented environment this may be interpreted as the ability to please the eye. In a performance-oriented environment it expresses the suitability in usage to complete a task.
As a quantity it may be associated in the mathematical sense with a functional, whose underlying dimensions are perceptual attributes. Multi-dimensional analysis of image quality judgement revealed in some cases dimensions which are close to what one would expect on the basis of common daily experience (Nakayama et al. 1980). In the case of TV images, global brightness, brightness contrast, sharpness and size are examples of such factors. In many cases these factors are the most direct link with the physical image parameters. However, sometimes the dominant dimensions are not easily defined. In this paper the latter will be disregarded.

Measures of quality or impairment and the strength of perceptual attributes

The most direct approach to the measurement of psychological quantities such as perceptual quality or the strenght of its dimensions is provided by psychology itself.
At least three types of measurements were found to be useful in this area:

1) <u>Scaling</u> in the sense of ordering sensations (e.g. Torgerson 1958, Krantz et al. 1971). It is a very important tool in image quality research, but is also controversial (e.g. Poulton 1977, Techtsoonian 1973). How to make an adequate choice from the different scaling techniques available is still a matter of great interest. This will be discussed in greater details below.

2) <u>Matching</u> is a procedure to equate stimuli subjectively with respect to some criterion (e.g. Bock and Jones 1968). In this context it is used to determine trade-off functions between different competing physical parameters or to check iso-sensation curves obtained from scaling experiments. Examples will be given later on.

3) <u>Thresholds</u> are usuaslly expressed in the strength of a physical parameter needed to make a perceptual attribute just visible (e.g. Laming 1973). In the context of image quality, thresholds can be viewed as either the upper limits of signals that introduce impairments or the lower limits of parameters which control attributes meant to be seen.

Within the limitation of this paper only a few examples can be given. A kind of serendipity is unavoidable.

Representativeness of scaled values

Scaling according to categories such as adjectives or numbers is one of the fastest methods to order sensations in rank. However, occasionally doubts concerning its validity occur. Questions are raised such as: Are the results obtained from a group of subjects representative for an other group? Are the results reproducible, especially in different laboratories? Are the results monotonuously and unbiased related to the sensations? Can anything be said about the relation between the strength of the sensation and its rating?

Interlabority tests performed by international organizations such as CCIR or by large project organizations such as Eureka '95 showed results that are representative and reproducible. The condition is that the tests should be performed with identical material under well controlled conditions. A trial series containing almost the complete range of the parameter set and preceeding the test helps the subjects to establish their internal scale. Fig. 1 shows the classic example of the COST 211 test. Four coding algorithms for videophone systems were compared on a category scale, using the adjectives given in Fig. 2 (Allnett et al. 1983). By means of Thurstones' model the scale was transformed into numbers (e.g. Torgerson 1958). Note that the results from the subjects of 5 different countries show a satisfactorily parallel. The judgements for the algorithms are quite different and demonstrate the sensitivity of the method and the effect of different scenes. The latter indicates the importance of a proper choice of standard scenes.

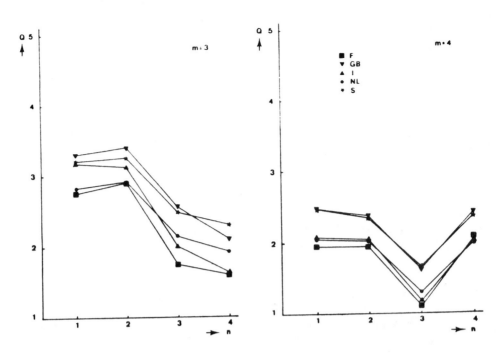

Fig. 1   Some results of the COST scaling experiment, performed
          in 5 European countries as indicated. A numerical
          quality scale derived from assessments according
          adjectives is plotted for 4 different algorithms.
          Every point represents the average of about 30
          subjects. The results of two test scenes are shown
          (m=3, m=4).

France	Great Britain	Italy	Netherlands	Sweden
Excellent	Excellent	Ottimo	Uitstekend	Utmarkt
Bon	Good	Buono	Goed	God
Assez bon	Fair	Discreto	Voldoende	Acceptabel
Médiocre	Poor	Scadente	Onvoldoende	Dålig
Mauvais	Bad	Pessimo	Slecht	Oanvündbar

Fig. 2   The adjectives of 5 European languages used to scale
          perceptual image quality plotted in Fig. 1.

This will be confirmed in the next example under TV conditions.
It concerns the effect of gamma on perceptual image quality. In
this case quality is assessed on a numerical 10-point category
scale. Generally, subjects handle this type of scale easily.
They stabilize their internal scale after having seen the
introductory trial series. The typical advantage of a numerical
category scale is its flexibility. The difficulty of finding
proper adjectives along one dimension especially in the case of
extremely small or wide ranges can be circumvented. The
subjective equidistancy of the numbers can be tested easily
(Edwards 1957). If necessary the scale can be transformed in an
equal interval scale by the method mentioned previously.

Intersubject – compared with interscene effects

Fig. 3 shows perceptual quality judgement of black and white TV images generated by an image processing system. The independent variable gamma is the power that characterizes the non-linear luminance reproduction curve of the imaging system, which is a power law in good approximation. The average luminance, being also a quality dimension, is kept constant here. Although subjects are consistent with one another in their judgement, different scenes show surprisingly large differences in the optimal value of gamma. In Fig. 4 the average values of all subjects are plotted, demonstrating the effect of scenes more clearly.

Brightness contrast, which is the dominant changing quality factor in changing gamma, was scaled by the same subjects and the same scenes using the same methods. The results are shown in Fig. 5. In Fig. 6 the quality judgements of Fig. 4 are plotted as a function of the rating of perceived brightness contrast of Fig. 5. All curves tend to coincide, suggesting that the reason for the different optimal values of gamma in Fig. 4 is closely connected with different transformation of gamma in brightness contrast for the various scenes. We will not discuss this in greater detail here.

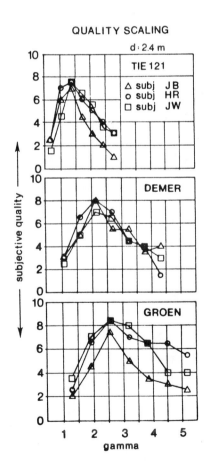

Fig. 3 Quality ratings by three subjects as a function of the (overall) values of gamma for three different scenes. Every scene has a constant mean luminance (TIE 121, 7.5 cd.m^{-2}; DEMER, 29.3 cd.m^{-2}; GROEN, 13.9 cd.m^{-2}).

Numerical category scales versus non-metric scales

Comparison of pairs is a non-metric type of scaling (Shepard

195

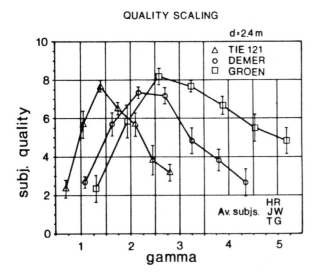

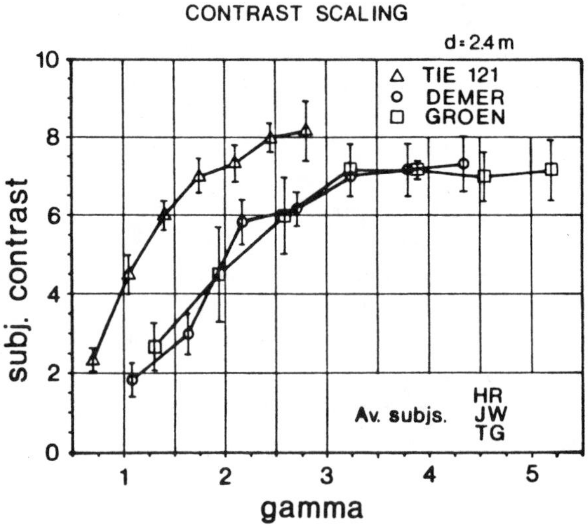

Fig. 4　Mean quality rating by the three subjects in figure 3 as a function of gamma for the three scenes indicated.

Fig. 5　Mean values of (subjective) global brightness contrast scaled by the subjects in figure 3 as a function of gamma for the same scenes, of which the mean luminance was kept constant.

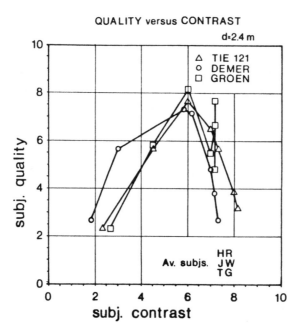

Fig. 6　Mean scaled image quality taken from figure 4 as a function of the scaled brightness contrast in figure 5.

1966, Kruskal and Wish 1978). Subjects are not compelled to handle numbers. Merely distances or similarity of two pairs are ordered. This method is considered to have a high validity, however it is also laborious. Fig. 7 shows a comparison of a numerical 10-point scale with the result of the comparison of pairs. In this case subjective sharpness, another important quality dimension (Nakayama et al. ibid) was scaled. The independent variable is the 6dB cut-off frequency of a second

order 2D spatial filter. The pointspread function of this filter
is convolved with the image in order to change spatial resolution of the image in a controlled way. The parameter is gamma,
described above. Since the non-metric experiment was very time
consuming, it was only done by one subject, while the category
scaling results are the averages of 3 subjects. The results
obtained with the numerical category scale are very similar with
those of the non-metric scale. Of course, this does not imply
the generalization that the category scale is as valid as the
non-metric scaling all circumstances.

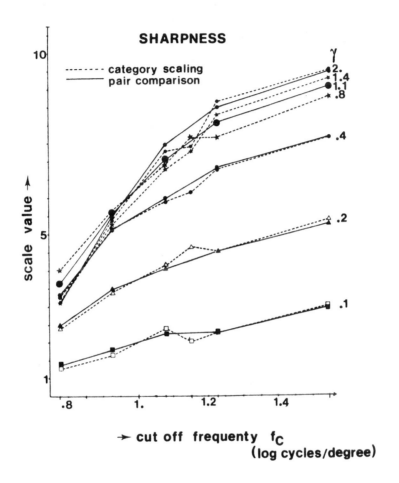

Fig. 7   Scaled perceived sharpness as a function of the 2D
          spatial cut-off frequency, gamma being the parameter.
          The category scale values are an average of 4 subjects
          RB, TL, MB and PS. The pair comparison scale values
          are the data of subject PS.

Consistency of matching and scaling

The fact that two dimensions are involved allows us to construct
trade-off functions, since horizontal iso-sharpness lines
intersect the family of curves at different cut-off frequency-
gamma combinations. These iso-sharpness trade-off functions are
shown in Fig. 8. This brings us to the possibility of checking
these curves by direct matching. Subjects appeared to be able to
match two images of an identical scene with different resolution
by varying gamma. These resuls are also shown in Fig. 8, and are
generally consistent with the iso-sharpness curves. It is of
some interest to see whether such a consistency would also be
the case if a more global attribute such as perceptual quality

is judged. In Fig. 9 the perceptual quality of projected images of different luminances is rated as a function of image size (Van der Zee and Boesten 1980). An adjective category scale was used, which was translated into a number scale by Thurstone's model. Since two variables are involved, iso-quality trade-off functions can be constructed. In Fig. 10 these functions are drawn and compared with matching results obtained with identical images of different sizes varying the luminance (Boesten and Van der Zee 1981). In view of the complexity of the attribute perceptual quality, the consistency between these results is quite encouraging.

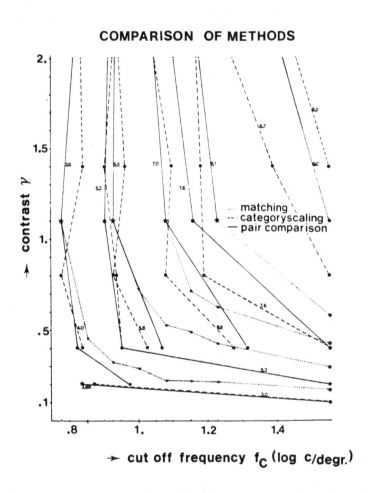

Fig. 8   Gamma versus cut-off frequency trade-off functions derived from the iso-quality curves of numerical category scaling and of the pair comparisons. Direct matching is also drawn in the same plot.

Impairments

Sometimes it is more appropriate to scale impairment instead of quality (De Ridder and Majoor 1988). This is, for instance, the case if one is dealing with impairments due to quantization errors in certain image coding algorithms. In Fig. 11 a simplified diagram of scale-space coding (Martens and Majoor 1989) is drawn.
Only two layers of binomial kernels with the two smallest dimensions are shown. In the first layer the original image is sampled by an array of 2D kernels which are 2 pixels apart. The resulting sampled image can be interpolated and compared with

the original. The prediction error signal is transmitted. In the next layer this procedure is repeated and so on. Important quantities are the quantization steps q_0 and q_1 of the quantizers Q. Large steps cause image errors which become manifest subjectively as unsharpness and speckles.

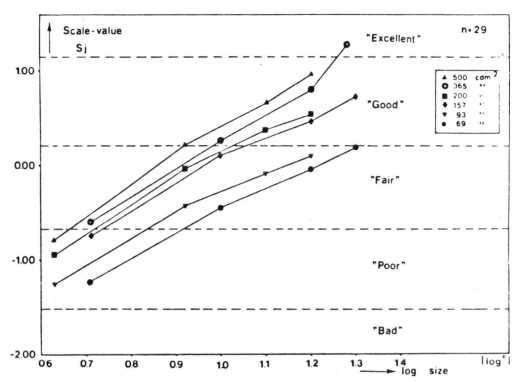

Fig. 9   Perceptual image quality of projected images, rated by an adjective category scale and translated into numbers is plotted as a function of size. The opengate luminance is the parameter. The points are averages of 29 subjects and 5 scenes (Van der Zee, Boesten 1980).

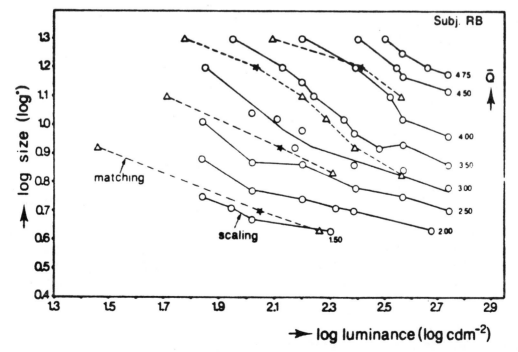

Fig. 10   Log size versus log luminance, as derived from the iso-quality curves of figure 9. The direct matching of one of the subjects has been drawn for comparison (Boesten, Van der Zee 1981).

199

Fig. 12 shows the results of perceived impairment, using a numerical category scale, as a function of the step size q_0, in the case where only the signal of the finest mesh is quantized.

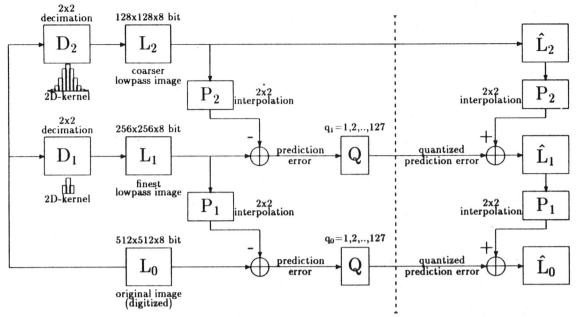

Fig. 11   Simplified diagram of a scale-space coding algorithm (Martens, Majoor). The effect of the quantization of the error signal q_0 or q_0 and q_1 was studied.

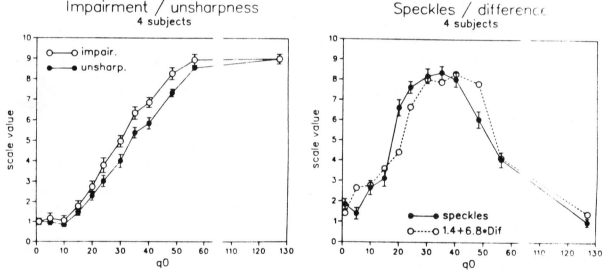

Fig. 12   Impairment as a function of the quantization step q_0 of the error signal of the finest low-pass mesh. The same subjects rated also the unsharpness, being one of the impairment dimensions.

Fig. 13   Solid line: scaled values of the perception of speckles of the images used in figure 12. Dashed line: A linear combination of the difference between the curves of figure 12.

In the same figure the strength of unsharpness, being one of the impairment dimensions, is plotted. The scale values of impairment and unsharpness were equalized at the largest quantization step because all subjects reported that at this step the impairment consisted of unsharpness only. The data demonstrate a systematic difference in functional behaviour between the scaled impairment and the scaled sharpness. A linear transformation of the difference is plotted in Fig. 13, together with the rating of the perception of speckles. The similarity of the curves suggests additivity of the two impairment dimensions in accordance with Alnetts (1983) assumption.

At threshold level it is of some interest to see how quantization errors produced in two layers of different mesh widths combine. Fig. 14 shows thresholds of various combinations of q_0 and q_1. The curves indicate that the thresholds of the two quantization errors are almost independent. (At supra threshold level, however, this is not true, as will be shown elsewhere.) Once more the scene does affect the relations relatively strongly, although the independence is conserved.

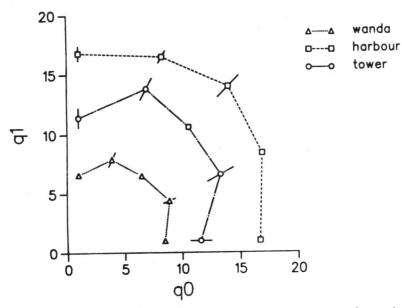

subject: HR

Fig. 14   Threshold of impairment due to quantization errors caused by a combination of quantization errors caused by a combination of quantization steps in the finest and the next layer schematically given in figure 11. The effect on the results of different scenes is demonstrated.

Expert and non-expert observers

At this point it seems appropriate to mention the expert-non-expert problem. It is well known that experts are considerable more critical than non-experts. However, in practice developments often take place on the basis of expert observation. In many cases the experts are not only experienced viewers, they also know where to look as a result of their knowledge of the equipment in addition, they tend to make a judgement on the basis of a scene in which they know the exact location of the sensitive spots. This is particularly the case at thresholds. The order of magnitude of these effects may be demonstrated in Fig. 15. This figure concerns thresholds of coding errors resulting from an algorithm for digital video recording (Westerink 1989). Two images were presented to the subjects, one being the original the other the coded. The subjects had to express their preference in a 'Two Alternative Forced Choice' procedure.

The independent variable is the bit rate used by the encoder (the more bits, the more detailed the image reconstruction). Five expert viewers were compared with five expert viewers with advance knowledge. For comparison the averages of two non-experts were included. The results show that experts who know what to look for and where are extraordinary critical. Generally speaking, it is necessary to decide on what type of subjects the conclusions are to be based.

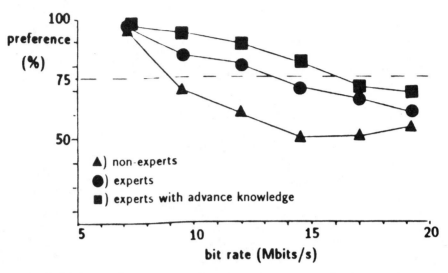

Fig. 15 Probability of preference of the original image over the coded image determined by Two Alternative Forced Choice. The bit rate of the encoder is the independent variable.

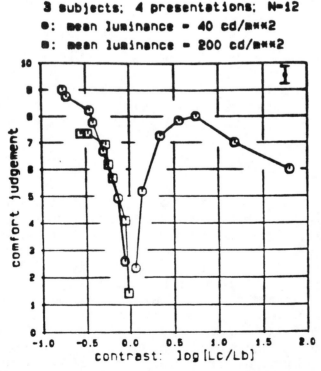

Fig. 16A The means of comfort judgements of 3 trained subjects as a function of log contrast ratio L_c = character luminance, L_b = background luminance.

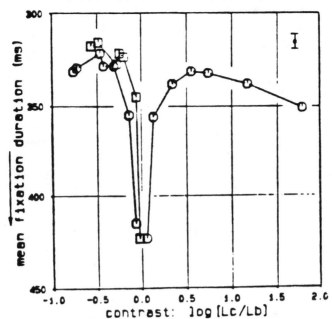

Fig. 16B Mean fixation duration (plotted downwards) for the same subjects as figure 2, plotted against log contrast ratio.

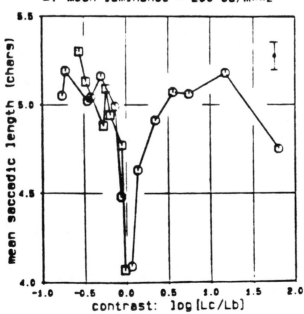

Fig. 16C Mean saccadic length (in number of characters) as a function of log contrast ratio. Same subjects as figure 2.

Performance measures

Finally an example is taken from a performance-oriented environment; namely VDUs.

If one assumes that reading comfort is a good criterion for the perceptual quality of VDUs, scaled comfort can be compared with some objective variables. From reading research it is known that, if a text is difficult, the saccades of the eye movements are relatively small and the fixation duration large. By taking a nonsense text and giving the subject a search task, for instance look for character A, the same observation is made if the visual quality of the characters is poor (Roufs et al. 1988). Also the scan velocity expressed in characters per second, calculated from the time needed to complete the task, becomes lower when the quality of the characters is made worse (Roufs et al. ibid). One way to vary the perceptual quality is to vary the luminance contrast ratio of character and background.

Fig. 16 shows a comparison of visual comfort scaled on a 10-point numerical category scale with the three objective variables mentioned. The data are averages of 3 subjects for the character font BEEHIVE. Fixation duration is plotted downwards, since it becomes longer if comfort decreases.

The correlation between the variables is obvious (the correlation coefficient $r=.87$).

Conclusions

The results may be summarized as follows:
a)   Scaling is a realistic tool to establish the relation between perceptual quality and the physical parameters.
b)   Numerical category scaling is fast and the results are encouraging with respect to its validicy.
c)   Matching is a viable method for the construction of trade-off functions and to test scaling results.
d)   Threshold measurement, a well-accepted method, is a useful tool to establish upper limits of unwanted and lower limits of wanted attributes.
e)   An appropriate choice of the test scenes causes more problems than the differences between subjects.
f)   Some caution has to be taken with test results measured by expert subjects, especially those who have advance knowledge.

References

Allnatt, J.W.; Gleis, N.; Kretz, F.; Sciarappa, A; Van der Zee, E. (1983)
    Definition and validation of methods for subjective assessment of visual telephone picture quality CSE2T. Rapporti technici XI, 59-65.
Allnatt, J.W. (1983)
    Transmitted picture assessment.
    John Wiley and Sons, N.Y.
Bock, RD and Jones, L.V. (1968)
    The measurement and prediction of judgement and choice.
    Holden Day, San Francisco.
Boesten, M.H.W.A.; Van der Zee, E. (1981)
    Psychophysical versus psychometric methods in image quality

measurements.
IPO Ann. Progr. Rep. 16, 67-71.

De Ridder, H.; Majoor, G.M.M. (1988)
Subjective assessment of impairment in scale-space coded images.
IPO Ann. Progr. Rep. 23, 55-64.

Edwards, A.L. (1957)
Techniques of attitude scale construction
Appleton Century Crofts Inc., N.Y.

Krantz, D.H.; Luce, R.D.; Suppes, P.; Tversky, A.
Foundations of measurement.
Acad.Press, N.Y.

Kruskal, J.B. and Wish, M. (1978)
Multidimensional scaling.
Sage Publications, London.

Laming, D. (1973)
Mathematical psychology.
Acad. Press, London.

Nakayama, T.; Massaak, K.; Honjyo, K.; Nishimito, K. (1980)
Evaluation and prediction of displayed image quality.
Proc. of SID 80, Digest 180-181.

Martens, J.B.O.S.; Majoor, G.M.M. (1989)
The perceptual relevance of scale-space image coding.
Signal Processing, August 1989.

Poulton, E.C. (1977)
Quantitative Subjective Assessments are almost always biased, sometimes misleading.
Biol. J. Psychol. 68, 409-425.

Roufs, J.A.J.; Boschman, M.C.; Leermakers, M.A.M. (1988)
Visual comfort as a criterion for designing display units.
In: Human-computer Interaction: Psychonomic Aspects.
Eds. G.C. van der Veer and G. Mulder, Springer-Verlag, 54-74.

Roufs, J.A.J.; Bouma, H. (1980)
Towards linking perception research and image quality.
Proc. SID 21, 247-270.

Shepard, R.N. (1966)
Metric structures in ordinal data.
Journal of Math. Psychol. 3, 287-315.

Teghtsoonian, R. (1973)
Range effects in psychophysical scaling and a revision of Stevens law.
Am. Journal of Psychol. 86, 3-27.

Torgerson, W.S. (1958)
Theory and methods of scaling.
John Weley & Sons Inc.

Van der Zee, E.; Boesten, M.H.W.A. (1980)
The influence of luminance and size on the image quality of complex scenes.
IPO Ann. Progr. Rep. 15, 69-75.

Westerink, J.H.D.M. (1989)
Influences of subject expertise on quality assessment of digitally coded images.
SID 89 Digest, 124-127.

TECHNICAL PROGRAM COMMITTEE

James Larimer, *Conference Chair*
NASA Ames Research Center

William E. Glenn
Florida Atlantic University

Walter Makous
University of Rochester

Joyce Farrell
Hewlett Packard